FROM PYTHAGORAS
TO EINSTEIN

NEW MATHEMATICAL LIBRARY

PUBLISHED BY

THE MATHEMATICAL ASSOCIATION OF AMERICA

The New Mathematical Library (NML) was begun in 1961 by the School Mathematics Study Group to make available to high school students short expository books on various topics not usually covered in the high school syllabus. In a decade the NML matured into a steadily growing series of some twenty titles of interest not only to the originally intended audience, but to college students and teachers at all levels. Previously published by Random House and L. W. Singer, the NML became a publication series of the Mathematical Association of America (MAA) in 1975. Under the auspices of the MAA the NML will continue to grow and will remain dedicated to its original and expanded purposes.

FROM PYTHAGORAS TO EINSTEIN

by

K. O. Friedrichs

Courant Institute of Mathematical Sciences

New York University

16

MATHEMATICAL ASSOCIATION

OF AMERICA

Acknowledgments

In writing this book I benefited greatly from the constructive criticism of those who have read the manuscript: colleagues and friends at the Courant Institute of Mathematical Sciences and the Rockefeller Institute, Lipman Bers who suggested the title, members of the Editorial Panel of the New Mathematical Library, graduate, undergraduate, and high school students Greg Chaitin, Melvin Friedman, Christopher Friedrichs, Francis Schrag, Leopoldo Toralbella, and Steven Zink. My thanks are due to all of them. I am particularly grateful to Anneli Lax for her untiring efforts and her encouragement.

K. O. F.

The editors of the New Mathematical Library also wish to acknowledge the generous help given by the many high school teachers and students who assisted in the preparation of this monograph. They are interested in reactions to the books in this series and hope that readers will write to: Anneli Lax, Editor, New Mathematical Library, New York University, The Courant Institute of Mathematical Sciences, 251 Mercer Street, New York, N.Y. 10012.

The Editors

Illustrated by George H. Buehler

Eighth Printing

Library of Congress Catalog Card Number 65-24963

Complete Set ISBN-0-88385-600-X
Vol. 16 ISBN-0-88385-616-6

Manufactured in the United States of America

Contents

NEW MATHEMATICAL LIBRARY

Other titles in preparation.

Preface

The present book is not addressed to a well defined group of readers. The first chapter is based on a lecture given to a special mathematics class of the sixth grade. The material of the second chapter belongs to elementary algebra; that of the third, fourth and fifth chapters may be studied by students of the twelfth grade. The subject of the sixth and seventh chapters may be accessible to selected high school seniors, but might just as well be read by college seniors. Knowledge of elementary Euclidean geometry is presupposed, and some familiarity with the basic notions of physics will be helpful.

It is hoped that this book will also be useful to teachers as a source of material and points of view not covered in the regular curriculum. The main viewpoints and guiding ideas in this book are explained in the introduction.

In order not to interrupt the unity of the presentation, many details have been skipped; the filling in of such details may perhaps serve in lieu of exercises.

The subject of this book is dominated by a definite theme strung along a somewhat meandering thread. Naturally, the author would be happy if there were at least one reader who would follow this thread from beginning to end.

To
Walter, Liska, David,
Christopher, and Martin

Introduction[*]

The aim of the present exposition is to discuss the Pythagorean theorem and the basic facts of vector geometry in a variety of mathematical and physical contexts, and to point out the significance of these notions in the special theory of relativity.

The Pythagorean theorem has suffered the same fate that so many basic mathematical facts have suffered in the course of the history of mathematics. At first, these facts were surprising when they were discovered and deep in that they required original inventive proofs. In the course of time such facts were placed into a conceptual framework in which they could be derived by more or less routine deductions; finally, in a new axiomatic arrangement of this framework, these facts were reduced to serve simply as definitions. Still, this need not have meant reduction to insignificance. What had become merely a definition may have been brought alive and made effective as a guiding principle in the development of new branches of mathematics. It is one of our aims to show that just this process describes the life cycle of the Pythagorean theorem.

In the first chapter of this exposition we begin by discussing one of the simplest proofs of the Pythagorean theorem within the framework of Euclidean geometry, and then we present a less frequently used proof. The latter is based on the fact that it is possible to plaster the plane by the small and the medium-sized squares of the Pythagorean figure as well as by the large squares, and that both coverings

* This introduction is meant primarily for those more or less familiar with the material covered in this book.

are reproduced under the same translation. We have selected this proof because of its striking intuitive appeal and the strong contrast it offers to the "second stage" proof discussed later on.

In the second chapter, under the heading of "signed numbers", we give a brief resume of the basic rules of operating with positive and negative real numbers to prepare the ground for the introduction of the notion of vector in the third chapter.

The notion of vector shares (with many other mathematical notions) a kind of fate different from that suffered by the Pythagorean theorem. This notion is fundamental to a large segment of mathematics and, in particular, of mathematical physics; nevertheless, it has not gained entrance into the teaching of geometry at an early stage. One may wonder whether the resistance offered to accepting this notion is of the inertial variety, or whether it is due to the suspected difficulty of relatively abstract new mathematical concepts.

After the rules of operation with vectors have been established, a routine application of these rules automatically leads to the Pythagorean theorem. This then is our "second stage" proof.

The third stage in the life cycle of this theorem will appear after we have discussed the notion of components of a vector with respect to a system of orthogonal unit vectors. We shall show how advantageous it is to change one's attitude and to define vectors as entities given by their components; for then the rules of operation are immediately obvious. But in this switch of attitude the Pythagorean statement ceases to be a theorem; it is employed as a definition. Still, the Pythagorean statement couched as definition retains its significance: it becomes the guiding principle in the development of geometry in more than three, in fact, in infinitely many dimensions. This development has, in many ways, provided the appropriate framework of the mathematical description of nature.

In the second part of the present exposition we shall present some fundamental principles of mechanics in which vectors and vector operations are used as basic tools. We shall not elaborate on the notion of force as vector, but confine ourselves to considering velocity and momentum as vectors.

The concept of momentum plays its major part in the theory of impact. In this process two moving particles hit each other and then either move away from each other in different directions or form a "compound particle". To be sure, these processes of "elastic" and "inelastic" impact belong to the basic phenomena of physics. In a sense, impact is the basic process involved in the interaction of elementary particles.

Accepting the laws of conservation of momentum and energy as fundamental, we shall deal first with elastic impact. In dealing with inelastic impact we take the opportunity to discuss the notions of thermal and of potential energy (without using the notion of work) as possible "internal energies" of the compound particle maintaining conservation of energy. Inelastic impact in reverse will be called an "explosion process" and illustrated with the action of a gun and a rocket.

The final aim of this exposition is to show the roles of the concept of vector and of the Pythagorean theorem (in a modified form) in the theory of relativity. Our major concern will again be with the problem of impact; but at first we shall give an account of the basic propositions of the special theory of relativity.

To explain our attitude towards the special theory of relativity, we add a few remarks addressed to those familiar with this theory.

We shall emphasize that the contentions of the kinematic part of this theory concern the behavior of rigid rods and clocks (i.e. springs) in motion, and not the nature of space and time as such. These contentions can be formulated most naturally as statements about space and time measurements made with the aid of rods and clocks. These statements in turn can be expressed concisely by reference to a vector geometry which employs an inner product associated with an indefinite metric form. It is at this place that the Pythagorean theorem reappears after having undergone a radical metamorphosis.

In describing Einstein's kinematics in terms of rods and clocks we shall not refer to propagation of light; the identification in this kinematics of the speed c with the speed of light will be relegated to additional considerations.

Passing over to mechanics in the theory of relativity, we shall employ as natural analogue of the momentum vector in classical mechanics a four-dimensional vector whose fourth component is closely related to the kinetic energy augmented by a term mc^2, the "rest energy", which at that stage cannot yet be given a physical interpretation. We shall use this energy-momentum vector to formulate the laws of elastic and inelastic impact. It is in connection with inelastic impact in reverse that we shall be led to Einstein's interpretation of the rest energy.

It will be evident that this exposition is not guided by "singleness of purpose". Different threads of thought of varying significance are intertwined. The level of sophistication rises quickly. Nevertheless, it is hoped that this exposition may be helpful to those who desire to enter the rich and colorful world of mathematics and mathematical physics.

CHAPTER ONE

The Pythagorean Theorem

The theorem ascribed to Pythagoras is concerned with the sides of a right triangle. The three sides of such a triangle are the two legs adjacent to the right angle and the hypotenuse opposite to this angle. The Pythagorean theorem gives a relation between the lengths of the three sides; it enables one to compute the length of the hypotenuse if the lengths of the other sides are given.

To determine the length of a side, or of any segment of a straight line, one must adopt a unit, or rather a particular segment whose length is taken as unit. Then one may express the length of any segment as a multiple of such a unit segment. To say the length of a segment is three means that it is three times as long as the unit segment.

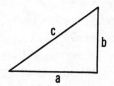

Figure 1. Right triangle with legs a, b and hypotenuse c

In this sense we denote the lengths of the two legs by a and b and the length of the hypotenuse by c (Figure 1). The Pythagorean theorem is embodied in the formula

$$c^2 = a^2 + b^2.$$

Clearly, when the lengths a and b are given, the length c can be computed since one can compute the square root of any positive number. Thus, by taking the square root of each member of the formula above, we find

$$c = \sqrt{a^2 + b^2}.$$

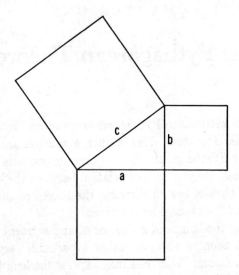

Figure 2. Right triangle with squares over legs and hypotenuse

The formula $c^2 = a^2 + b^2$ can be given a direct geometrical meaning. Let us erect the square over each of the two legs of the triangle and over its hypotenuse; see Figure 2. Clearly, the area of a square is the square* of the length of its side. The significance of the formula $c^2 = a^2 + b^2$ can therefore be expressed by saying that the area of the square over the hypotenuse is the sum of the areas of the squares erected over the sides. As a matter of fact, it is this geometrical assertion which is called Pythagoras' theorem by Euclid.

How can the Pythagorean theorem be proved? Many different proofs have been given and we shall present various quite different approaches leading to such proofs.

---

* Here the first word "square" refers to a figure, the second one to the product of a number by itself.

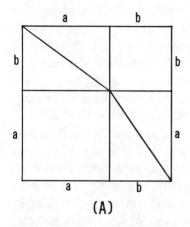
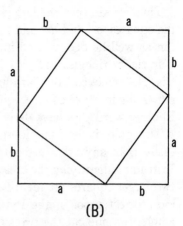

(A)   (B)

Figure 3. Proof of the Pythagorean theorem (A) by decomposing a square of area $(a + b)^2$ into two squares of areas a^2 and b^2 and four triangles of areas $\frac{1}{2} ab$; (B) by decomposing the same square into four triangles of areas $\frac{1}{2} ab$ and one square of area c^2

In the present chapter we shall give two geometrical proofs. In the first we consider the square with the side* $a + b$; see Figure 3. In one corner of this "large square" we place the square with the side* a ; in the opposite corner we place the square with the side b. Evidently two rectangles, each with sides a and b, remain. Thus the area of the large square is the sum of the areas a^2 and b^2 of the squares with sides a and b and twice the area ab of the rectangle with sides a and b; see Figure 3A.

Each rectangle may be regarded as the sum of two right triangles, each with legs a and b. Now, one may place these four triangles differently in the large square, as indicated in Figure 3B, so that each is placed in a corner of the large square with its short leg to the left and long leg to the right if looked at from the outside. The f.gure bounded by the hypotenuses of these triangles is evidently a quadrilateral each of whose sides has length c. In fact this figure is a square; for, at a point of the side $a + b$ where two triangles meet, three angles combine to form a straight angle. Two of them, being opposite to legs a and b of two congruent right triangles, are complementary and therefore add up to a right angle; so the remaining angle is also a right angle.

---

* Here we use the term "side" where we should have used the expression "side with length". This is convenient and we shall continue to use this convenient simplification of language.

Thus we see that the large square, which can be decomposed into the squares a^2 and b^2 and four right triangles with legs a and b, can as well be decomposed into the square with area c^2 and again four right triangles with legs a and b. Imagining the four right triangles removed from both figures, we realize that the area $a^2 + b^2$ remaining in the one figure equals the area c^2 remaining in the other. In other words, we have derived the statement $a^2 + b^2 = c^2$.

This proof is very simple—perhaps it is grasped by intuition more easily than any other geometrical proof—but it is not quite direct. Instead of identifying the areas $a^2 + b^2$ and c^2 directly, the areas of two larger squares are identified. One may wonder whether or not a more direct proof of the Pythagorean theorem is possible. For example, can one cut the two squares a^2 and b^2 into pieces and then recombine these pieces to form the square c^2? Such a dissection and recombination is indeed possible; as a matter of fact, it is possible in infinitely many ways. We present a simple "recipe" for this purpose.

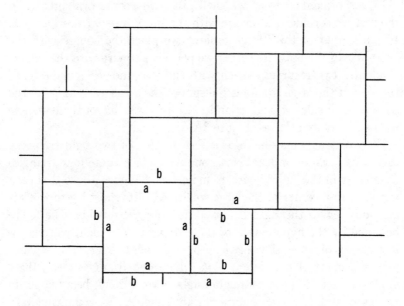

Figure 4A. Covering of plane by squares with sides a and b

We imagine the whole plane covered by squares with sides a and b in the manner shown in Figure 4A; we also imagine the plane covered by squares with side c in the manner shown in Figure 4B. We shall

refer to these coverings as the first and second "arrays". The sides
of the second array are to be parallel to the hypotenuses of the right
triangles whose legs a and b are so placed that they are parallel or
perpendicular to the sides of the squares of the first array.

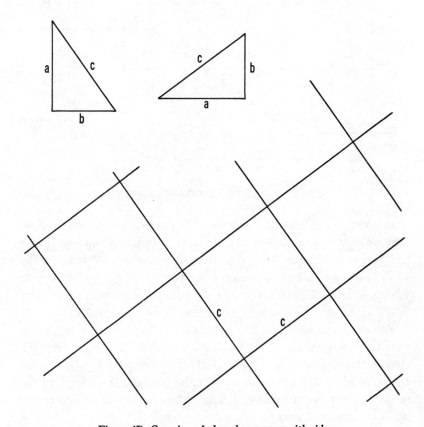

Figure 4B. Covering of plane by squares with side c

There are many possibilities for the location of the second array
in relation to the first one. We certainly can choose the location of one
of its vertices at pleasure; of course, the location of all other vertices is
then fixed (see Figure 5). If the selected vertex of the second array
lies in a particular square of the first array, all other vertices of the
second array lie in corresponding positions in congruent squares.

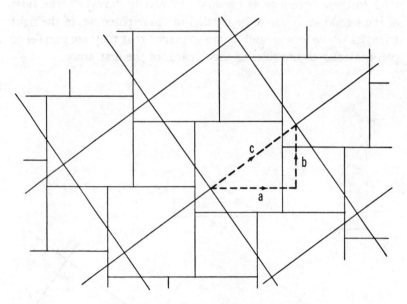

Figure 5. Second array superimposed on first array

In order to explain what we mean by "corresponding position" we observe that any square of the first array can be displaced into a congruent one simply by first moving it a distance a horizontally to the right and then a distance b vertically up, as is evident from Figure 5. In this displacement every point of the square goes over into a point in a corresponding position in the new square.

The displacement of each point may also be described by saying that the point moves along the legs a and b of a right triangle; clearly, this displacement can be effected more directly just by moving the point along the hypotenuse c of the triangle. If a vertex of the second array is moved along such a hypotenuse, it will evidently end up at a vertex of a congruent square.

Applying the same kind of argument to all four directions leading from a vertex of the second array to neighboring vertices, and repeating this any number of times, we realize that all vertices of the second array have corresponding positions in congruent squares of the first array.

Now, consider a pair of adjacent squares of the first array, say α, β with sides a and b, and a square $\mathcal{C}$ of the second array intersecting the squares α and β ; see Figure 6. Clearly, the sides of the squares of the second array cut the squares α and $\mathcal{B}$ into pieces. Each such piece either belongs to $\mathcal{C}$ or it belongs to one of the squares congruent to $\mathcal{C}$. In the latter case this piece is congruent to a piece lying within $\mathcal{C}$ cut out by the lines of the first array. If now all the pieces of α and $\mathcal{B}$ are cut out and reassembled within $\mathcal{C}$, each piece being placed in its corresponding position, then they completely fill out the square $\mathcal{C}$. Thus we recognize that the area of $\mathcal{C}$ is precisely equal to the sum of the areas of α and $\mathcal{B}$ and are led to the statement $c^2 = a^2 + b^2$, that is to say, to the statement of the Pythagorean theorem.

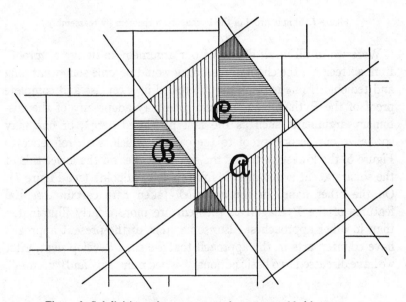

Figure 6. Subdivisions of squares α and $\mathcal{B}$ reassembled in square $\mathcal{C}$

To be sure, our present argument is not complete inasmuch as we have taken for granted a number of simple geometrical facts without deriving them rigorously. It would not be difficult to supply these missing links for each way of cutting up and reassembling our squares. If this is done, a proof of the Pythagorean theorem results; however, we have not carried out these details but have given only a "recipe" for proving the theorem. Note that we have given infinitely many such "recipes", since we can choose any one of infinitely many places for a vertex of the second array and hence can select the location of the square $\mathcal{C}$ from an infinity of possibilities.

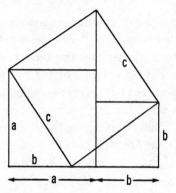

Figure 7. Simple proof of the Pythagorean theorem by reassembly

Were we to fill in all details of our argument to derive a "proof" from a "recipe", the chain of reasoning would become somewhat long and tedious. If one is primarily interested in a concise and complete proof of the Pythagorean theorem, one may adopt any of the customary arguments such as the first presented here*; or one may specialize our second proof to the case in which, with reference to Figure 6, the lowest vertex of the square c lies on the lower side of the square a at a distance b from its left endpoint (see Figure 7). On the other hand, in the approach taken here certain essential features of the Pythagorean theorem are more clearly illuminated than in other approaches. Also, some steps of the present approach have counterparts in the approach that we shall develop next, after we have discussed two basic notions: "signed numbers" and "vectors".

* or that indicated in the footnote of p. 28.

CHAPTER TWO

Signed Numbers

Signed numbers are introduced in algebra in order to enable one to subtract any number from any other, no matter which of the two is larger. The remarkable fact is that this can be done in such a way that the formal rules of algebraic operations are preserved.

Let the symbols s_1, s_2, s_3, $\cdots$ stand for any positive numbers. Then the following "formal rules" involving addition ($+$) and multiplication ($\cdot$) are valid:

I $$s_2 + s_1 = s_1 + s_2$$

II $$(s_1 + s_2) + s_3 = s_1 + (s_2 + s_3)$$

III $$(s_1 + s_2) \cdot s_3 = s_1 \cdot s_3 + s_2 \cdot s_3$$

IV $$s_2 \cdot s_1 = s_1 \cdot s_2$$

V $$(s_1 \cdot s_2) \cdot s_3 = s_1 \cdot (s_2 \cdot s_3)$$

We note that these rules remain valid if we enlarge the class of positive numbers by "adjoining" the number zero to this class.

Negative numbers afford a further enlargement. To every positive number p one assigns its negative $-p$ and defines two operations—also called "addition" and "multiplication"—as follows:

$$(-p_1) + (-p_2) = -(p_1 + p_2),$$

$$(-p_1) + p_2 = p_2 - p_1 \quad \text{if } p_2 \geq p_1,$$

$$= -(p_1 - p_2) \quad \text{if } p_2 < p_1,$$

$$p_1 \cdot (-p_2) = -(p_1 \cdot p_2)$$

$$(-p_1) \cdot (-p_2) = p_1 \cdot p_2.$$

Also, one sets $(-p_1) + 0 = -p_1$, and $0 \cdot (-p_2) = 0$. It is well known that, with these definitions, the five rules enumerated above hold just as for ordinary addition and multiplication.

Subtraction can be defined for our enlarged class of numbers by reducing it to addition; for any positive number p one simply sets

$$s - p = s + (-p), \qquad s - (-p) = s + p,$$

where the number s may be positive, negative, or zero. Subtraction according to this definition agrees with subtraction in the customary sense if $p < s$. In any case, subtraction can always be carried out.

An important notion concerning directed or signed numbers is the *absolute value* $|s|$ or magnitude of such a number s defined by

$$|s| = s \quad \text{if } s \geq 0 \quad \text{and} \quad |s| = -s \text{ if } s < 0.$$

The important inequality

$$|s_1 + s_2| \leq |s_1| + |s_2|,$$

valid for any two directed numbers s_1, s_2, should be mentioned.

These matters, well known from elementary algebra, have been repeated here since they serve, to some degree, as a guide to the introduction of the notion of vector in the next chapter. A certain class of mathematical entities, "positive numbers", was extended to a larger class of entities, "positive numbers, negative numbers, and zero", in such a way that the formal laws of operation valid for the original entities remained valid for those of the enlarged class. To this end it was necessary to define the basic operations, addition and multiplication, in an appropriate manner for the entities of the enlarged class.

One may wonder whether or not the basic operations for negative numbers could not have been defined differently. Of course, one could quite arbitrarily have introduced other operations instead of those we did introduce, and in principle one could have given the names "addition" and "multiplication" to any other such operations. But to do so would have been quite pointless; for, the "formal rules" would then not be valid for these operations. In fact, as could be shown, the only operations on negative numbers for which the formal laws (including those involving subtraction and operations with 0, 1 and −1) are valid are just the ones we have described. It is for this reason that the operations described are generally accepted as "addition" and "multiplication" of negative numbers or, rather, of the class of entities consisting of positive numbers, negative numbers, and zero. The entities of this class are frequently called *signed* or *directed* numbers. We shall hereafter call them just "real" numbers as is the custom in advanced branches of mathematics.

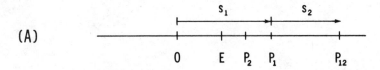

(A)

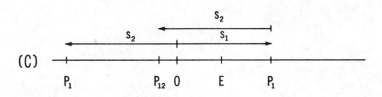

(B)

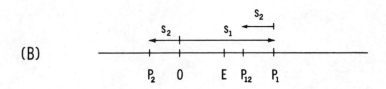

(C)

Figure 8. (A) Addition of positive numbers; (B) Addition of a positive number s_1 and a negative number s_2, where $-s_2 < s_1$; (C) Addition of a positive number s_1 and a negative number s_2, where $-s_2 > s_1$

The term "directed number" is suggested by a simple geometrical interpretation of the operation of addition. We consider a straight line, mark on it a point O, the origin, and a point E whose distance from O is taken as the unit of length. Every point P on the line can then be characterized by giving its distance from O and stating whether or not it lies on the same side of O as the point E. We assign to the point P a number s as follows: In case P is on the same side of O as E, we take s as its distance from O, i.e., $s = \overline{PO}$; in case P lies on the other side of O, we set $s = -\overline{PO}$; if P is O itself, we take $s = 0$. It is also convenient to introduce for every point P of the line the "arrow" $\overrightarrow{OP}$ which leads from O to P; if P is O, of course, this arrow shrinks to a point. Clearly, to every real number there corresponds such an arrow $\overrightarrow{OP}$ and vice versa. The addition of numbers s can now be interpreted as an operation with the corresponding arrows, as shown in Figures 8A, 8B, 8C.

CHAPTER THREE

Vectors

There are other classes of mathematical objects, besides the class of real numbers, for which operations can be introduced that obey essentially the same formal rules as ordinary addition and multiplication. We shall describe such a class of objects: vectors.

A "vector" is an entity characterized by a direction in space and a positive number called its "magnitude"; also, this entity is attached to a point in space. In the course of this exposition we shall meet various entities (such as the velocity of a moving particle) which are characterized by direction and magnitude, are attached to points, and hence may be regarded as vectors.

For the present we may assume that vectors are attached to the origin O unless otherwise stated. Moreover, we shall for now confine ourselves to describing a vector as an arrow leading from the "initial" point O to another point P in space, the "end point"; the magnitude of the vector is then taken to be the length of this arrow, i.e. the distance of P from O. From the different notations used by different authors to denote vectors, we select the notation $\bar{a}, \bar{b}, \cdots$ for vectors and $|\bar{a}|, |\bar{b}|, \cdots$ for their magnitudes.

We shall extend the notion of vector slightly by regarding as a vector the arrow which leads from the point O to itself. Never mind that nothing is left of this arrow but a point; it will do no harm that this "zero vector", denoted by $\bar{o}$, has no definite direction and that its magnitude is zero: $|\bar{o}| = 0$.

The first operation we introduce will obey exactly the same formal rules as ordinary addition; this operation is therefore called "addition of vectors", and the symbol $+$ is used to express it. To determine the sum $\vec{a} + \vec{b}$ of two vectors $\vec{a}$ and $\vec{b}$ (both attached to the same initial point O) we transplant the vector $\vec{b}$ parallel to itself so that its new initial point is the endpoint of $\vec{a}$. The arrow leading from O to the endpoint of the transplanted vector $\vec{b}$ is then defined to be the sum $\vec{a} + \vec{b}$; see Figure 9A.

Similarly, the vector $\vec{b} + \vec{a}$ is defined, after parallel translation of $\vec{a}$ to the endpoint of $\vec{b}$, as the arrow leading from O to the endpoint of the transplanted vector $\vec{a}$; see Figure 9B.

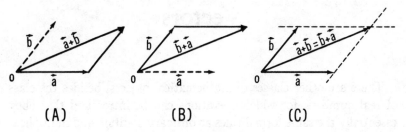

(A)   (B)   (C)

Figure 9. Addition of vectors

The construction of the vector $\vec{a} + \vec{b}$ could also have been described in the following manner: Through the endpoints of $\vec{a}$ and $\vec{b}$ one draws straight lines parallel to $\vec{b}$ and $\vec{a}$, respectively, and prolongs them until they intersect; see Figure 9C. Thereby, as is well known from elementary geometry, a parallelogram is formed. The sides opposite to $\vec{a}$ and $\vec{b}$ are parallel to $\vec{a}$ and $\vec{b}$ and have the same lengths as $\vec{a}$ and $\vec{b}$. Thus it is clear that the vector $\vec{a} + \vec{b}$ is just the arrow leading from the point O to the diagonally opposite point in the parallelogram. At the same time it becomes clear that, if in the original definition of addition the roles of $\vec{a}$ and $\vec{b}$ are interchanged, the result is the same. We express this fact by the formula

$$\text{I} \qquad\qquad \vec{b} + \vec{a} = \vec{a} + \vec{b}.$$

Evidently, this rule is the same as rule I for ordinary addition.

In the construction of the parallelogram we have tacitly assumed that the two vectors $\vec{a}$ and $\vec{b}$ are not parallel to each other; but even if they are, rule I is valid, as could be readily verified. (Note that the two vectors are assumed to be attached to the same initial point and hence lie on the same straight line if they are parallel.)

Rule II also carries over to vectors:

II $$(\bar{a} + \bar{b}) + \bar{c} = \bar{a} + (\bar{b} + \bar{c}).$$

To verify this rule, consider the three planes spanned by the three pairs of vectors $(\bar{a}, \bar{b})$, $(\bar{b}, \bar{c})$, $(\bar{c}, \bar{a})$, and pass three planes parallel to them through the endpoints of $\bar{c}$, $\bar{a}$, and $\bar{b}$ respectively. Thus a solid figure bounded by three pairs of parallelograms is formed; it is called a "parallelopiped"* (see Figure 10). Denote the arrow leading from O to the diagonally opposite point by $\bar{a} + \bar{b} + \bar{c}$. A simple consideration will show that this vector could be obtained by adding first two of the three vectors $\bar{a}$, $\bar{b}$, $\bar{c}$ and then the third one, in any order. Having verified this fact, one has verified the validity of rule II.

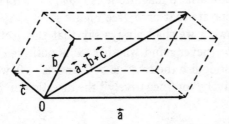

Figure 10. Addition of three vectors

In describing the formation of the parallelopiped we have tacitly assumed that the three vectors $\bar{a}$, $\bar{b}$, and $\bar{c}$ do not lie in a single plane. If they do, no parallelopiped is formed, properly speaking; instead, a figure is formed which lies completely in a plane—it might be called a "collapsed" parallelopiped. Nevertheless, the argument leading to rule II remains valid in this case. Figure 10 may be regarded as representing a collapsed as well as an uncollapsed parallelopiped.

* The parallelopiped is the analogue in space of the parallelogram in the plane. In the following we shall freely use a few simple basic facts of (three dimensional) space geometry.

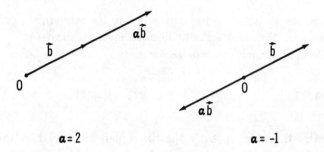

Figure 11. Multiplication of a vector by a positive number and by a negative number

Our next task, naturally, is to introduce the notion of *multiplication* of vectors. In fact, two different kinds of such "multiplication" will be described. The first one is the *multiplication of a vector* $\vec{b}$ *by a real number* α. To define it we first assume the number α to be positive: $\alpha > 0$. Then we define the vector $\alpha\vec{b}$ as the arrow which has the same direction as $\vec{b}$ but whose length is α times the length of $\vec{b}$: $|\alpha\vec{b}| = \alpha|\vec{b}|$ for $\alpha > 0$. If α is negative, $\alpha < 0$, we define $\alpha\vec{b}$ as the vector whose magnitude is $|\alpha\vec{b}| = -\alpha|\vec{b}|$, but whose direction is opposite that of $\vec{b}$; see Figure 11.

In the following we shall always say that the vector $\alpha\vec{b}$ has the direction of $\vec{b}$ (never mind whether α is positive or negative). If we want to distinguish the cases $\alpha > 0$ and $\alpha < 0$ we shall say that $\alpha\vec{b}$ has "positively" or "negatively" the direction of $\vec{b}$.

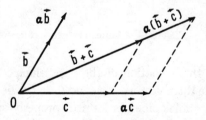

Figure 12. Multiplication of a sum of vectors

As a first analogue of rule III of Chapter 2 we formulate rule

III₁ $\alpha(\vec{b} + \vec{c}) = \alpha\vec{b} + \alpha\vec{c}$,

which is immediately verified from the law of proportionality; see Figure 12. There will be a second analogue of rule III and an analogue of rule V. Before formulating these we make a few general observations.

First of all we note that, corresponding to every vector $\bar{c}$, there is a "unit vector" $\bar{u}$, that is, a vector of length $|\bar{u}| = 1$, such that

$$\bar{c} = |\bar{c}|\,\bar{u}.$$

In case $|\bar{c}| \neq 0$, we need only set

$$\bar{u} = \frac{1}{|\bar{c}|}\,\bar{c};$$

if $|\bar{c}| = 0$ we may take any unit vector for $\bar{u}$. We shall refer to $\bar{u}$ as the unit vector (positively) in the direction of $\bar{c}$.

Next we observe that through the notion of vector one is quite naturally led to the interpretation of the real numbers as directed numbers in the manner described at the end of Chapter 2. One need only let each real number s correspond to the vector $s\bar{u}$, where $\bar{u}$ is the vector of length 1 leading from O to E.

From rules III and V for directed numbers one then derives the rules

III$_1'$ $\qquad\qquad\qquad (\alpha + \beta)\bar{c} = \alpha\bar{c} + \beta\bar{c},$

V$_1'$ $\qquad\qquad\qquad (\alpha\beta)\bar{c} = \alpha(\beta\bar{c})$

governing the multiplication of an arbitrary vector $\bar{c}$ by real numbers; for, writing the vector $\bar{c}$ in the form $\bar{c} = |\bar{c}|\,\bar{u}$ with $|\bar{u}| = 1$, we realize that rules III$_1'$ and V$_1'$ reduce to rules III and V of Chapter 2 if we identify $\alpha, \beta, |\bar{c}|$ with s_1, s_2, s_3.

We cannot give an analogue of rule IV for the multiplication of a vector by a number since we have not introduced multiplication of a vector by a number from the right.

Note that the result of multiplying a vector by a real number is a vector. In contrast to this multiplication we shall define a second kind of multiplication in which both factors are vectors, while their product is a real number. Before defining this kind of multiplication it is necessary to introduce the notion of "projection".

Let $\bar{a}$ and $\bar{c}$ be any two vectors. Then we may define the projection of $\bar{a}$ on $\bar{c}$ in the following manner: Through the endpoint of $\bar{a}$ we construct the plane perpendicular to the straight line which passes through O in the direction of $\bar{c}$, assuming $\bar{c} \neq \bar{o}$; see Figure 13. The arrow which leads from O to the point at which this line intersects the plane will be called the *projection* $\bar{a}_c$ of $\bar{a}$ on $\bar{c}$. (Of course, the endpoint of the vector $\bar{a}_c$ need not lie between O and the endpoint of $\bar{c}$; it need only lie on the line through $\bar{c}$.) In case $\bar{c} = \bar{o}$ we set $\bar{a}_c = \bar{o}$.

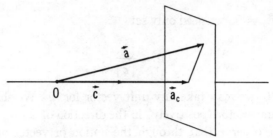

Figure 13. Projection of vector $\bar{a}$ on vector $\bar{c}$

Since the vector $\bar{a}_c$ is in the direction of $\bar{c}$, it is a multiple of the unit vector $\bar{u}$ having positively the direction of $\bar{c}$; that is to say, there is a number—we denote it by a_c—such that

$$\bar{a}_c = a_c \bar{u}.$$

Clearly, we have $|a_c| = |\bar{a}_c|$ for this number a_c. The term "projection", used to denote the vector $\bar{a}_c$, is frequently also used to denote the number a_c. The usage of this terminology is not definitely established. In case of doubt we shall distinguish between the projection-vector $\bar{a}_c$ and the projection number a_c.

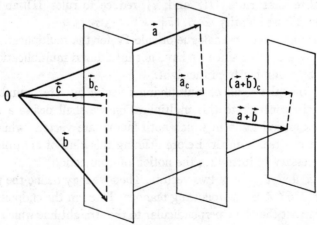

Figure 14. Projection of sum is sum of projections

Two formal rules will be established for projections. The first one is

$$(\bar{a} + \bar{b})_c = \bar{a}_c + \bar{b}_c.$$

To express this rule in geometric terms we construct the parallelogram generated by $\bar{a}$ and $\bar{b}$ and consider the three parallel planes perpendicular to $\bar{c}$ through the endpoints of $\bar{a}$, $\bar{b}$ and $\bar{a} + \bar{b}$ (see Figure 14). Our rule can then be stated by saying that transplanting the vector $\bar{b}$ to the endpoint of $\bar{a}$ and afterwards projecting the resulting vector $\bar{a} + \bar{b}$ on $\bar{c}$ yields the same result as first projecting $\bar{a}$ and $\bar{b}$ on $\bar{c}$ and then transplanting the projection $\bar{a}_c$ to the endpoint of $\bar{b}_c$. That this is so is readily deduced from basic facts of geometry.

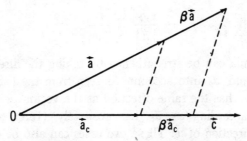

Figure 15. Projection of multiple of vector $\bar{a}$ is multiple of projection of $\bar{a}$

The second rule,

$$(\beta\bar{a})_c = \beta\bar{a}_c,$$

is an immediate consequence of the law of proportionality; see Figure 15.

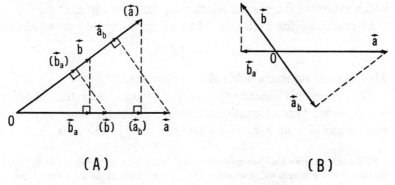

(A)   (B)

Figure 16. Proof of the relation $a_b/|\bar{a}| = b_a/|\bar{b}|$

The question arises naturally: what is the relationship between the projection of $\bar{a}$ on $\bar{b}$ and the projection of $\bar{b}$ on $\bar{a}$? This relationship can be read off from the two Figures 16 in which $\bar{a} \neq \bar{o}$ and $\bar{b} \neq \bar{o}$ is assumed. There the right triangle spanned by the vectors $\bar{a}$ and $\bar{a}_b$ is flipped around so that $\bar{a}$ goes over into a "new" vector $(\bar{a})$, which falls in the direction of $\bar{b}$, and $\bar{a}_b$ goes over into a "new" vector $(\bar{a}_b)$ in the direction of $\bar{b}_a$. It is then seen that the line connecting the endpoint of the new vector $(\bar{a})$ with the endpoint of the new vector $(\bar{a}_b)$ is parallel to the line connecting the endpoint of $\bar{b}$ with that of $\bar{b}_a$. Consequently, by virtue of the law of proportionality, the lengths of the vectors $(\bar{a}_b)$ and $(\bar{a})$ are in the same ratio as those of $\bar{b}_a$ and $\bar{b}$; that is to say,

$$\frac{|\bar{a}_b|}{|\bar{a}|} = \frac{|\bar{b}_a|}{|\bar{b}|}.$$

This formula can be strengthened by taking the directions of the vectors $\bar{b}_a$ and $\bar{a}_b$ into account. As seen from the two Figures 16, the vector $\bar{b}_a$ has the same direction as the vector $\bar{a}$, positively or negatively, according as $\bar{a}_b$ has positively (16A) or negatively (16B) the direction of $\bar{b}$. These two cases can also be distinguished by the signs of the projection numbers a_b and b_a, which are given by $a_b = |\bar{a}_b|$, $b_a = |\bar{b}_a|$ in the first case and by $a_b = - |\bar{a}_b|$, $b_a = - |\bar{b}_a|$ in the second case. Employing these projection numbers, we readily derive from the preceding formula the formula*

$$\frac{a_b}{|\bar{a}|} = \frac{b_a}{|\bar{b}|}$$

which expresses the desired relationship between $\bar{a}_b$ and $\bar{b}_a$.

After multiplying both sides of this formula by $|\bar{a}|\,|\bar{b}|$ we obtain

$$a_b\,|\bar{b}| = |\bar{a}|\,b_a.$$

This relation evidently holds also if $\bar{a} = \bar{o}$ or if $\bar{b} = \bar{o}$.

The expressions occurring in the last relation enter the definition of the second type of multiplication. We define the *inner product* of two vectors $\bar{a}$ and $\bar{b}$ to be the number $a_b\,|\bar{b}|$. The notation $\bar{a} \cdot \bar{b}$

---

* This formula may also be expressed as the statement that the cosine of the angle that the vector $\bar{a}$ makes with the vector $\bar{b}$ equals the cosine of the angle that $\bar{b}$ makes with $\bar{a}$, a statement we thus have proved.

is customary for this product. Thus

$$\bar{a} \cdot \bar{b} = a_b \,|\, \bar{b} \,| \,.$$

This definition applies also if $\bar{a} = \bar{o}$ or if $\bar{b} = \bar{o}$; clearly $\bar{a} \cdot \bar{b} = 0$ in such a case.

Because of the definition of $\bar{a} \cdot \bar{b}$, the relation $a_b \,|\, \bar{b} \,| \, = \, |\, \bar{a} \,| \, b_a$ derived above leads to the rule

IV$_2$ $$\bar{a} \cdot \bar{b} = \bar{b} \cdot \bar{a},$$

which expresses the fact that the order of inner multiplication can be interchanged. This rule is an analogue of rule IV for ordinary numbers; see Chapter 2.

Rule III also carries over to the inner multiplication of two vectors; it reads

III$_2$ $$(\bar{a} + \bar{b}) \cdot \bar{c} = \bar{a} \cdot \bar{c} + \bar{b} \cdot \bar{c}.$$

To prove its validity we refer to the relation

$$(\bar{a} + \bar{b})_c = \bar{a}_c + \bar{b}_q$$

derived earlier for the projections of the vectors $\bar{a}, \bar{b},$ and $\bar{a} + \bar{b}$ on the vector $\bar{c}$. In terms of projection numbers this formula can be written as

$$(a + b)_c = a_c + b_c,$$

where $(a + b)_c$ stands for the projection number of $\bar{a} + \bar{b}$ on $\bar{c}$. Multiplication of the terms of the last formula by $|\, \bar{c} \,|$ yields the formula

$$(a + b)_c \,|\, \bar{c} \,| = a_c \,|\, \bar{c} \,| + b_c \,|\, \bar{c} \,| \,,$$

which is the same as formula III$_2$, by definition of the inner product.

By virtue of IV$_2$, rule III$_2$ is equivalent to

III$'_2$ $$\bar{c} \cdot (\bar{a} + \bar{b}) = \bar{c} \cdot \bar{a} + \bar{c} \cdot \bar{b}.$$

The rather obvious rule

V$_2$ $$\alpha(\bar{b} \cdot \bar{c}) = (\alpha \bar{b}) \cdot \bar{c}$$

may be regarded as an analogue of rule V of Chapter 2. Naturally there is no analogue of rule V involving three vectors since the inner product of two vectors is not a vector but just a real number.

Incidentally, a real number is also called a "scalar" when it is to be contrasted with a vector; inner multiplication is therefore also referred to as "scalar multiplication".

Having established various algebraic rules for vectors we can demonstrate numerous geometrical facts simply by applying these rules formally. Some of these facts involve the notions of "length" of a segment and of "perpendicularity" of two segments. Both these notions can be described with the aid of the concept of inner product.

To give these descriptions we first consider the inner product of a vector $\bar{a}$ with itself. Clearly, the projection of $\bar{a}$ on $\bar{a}$ is $\bar{a}$ itself; for, the plane through the endpoint of $\bar{a}$ perpendicular to the line through $\bar{a}$ intersects this line at the endpoint of $\bar{a}$. From the relation $\bar{a}_a = \bar{a}$, which expresses this fact, we immediately obtain the formula

$$a_a = |\bar{a}| \; ;$$

this implies that the inner product of $\bar{a}$ with itself—given by $a_a |\bar{a}|$ according to definition—is the square of the magnitude of $\bar{a}$:

$$\bar{a} \cdot \bar{a} = |\bar{a}|^2.$$

Since the length of a segment may be regarded as the magnitude of the vector leading from one endpoint of the segment to the other, we now realize that this length can be expressed as the square root of the inner product of this vector with itself.

Next we consider two vectors $\bar{a}$ and $\bar{b}$ (attached to O) whose inner product vanishes: $\bar{a} \cdot \bar{b} = 0$. Then, unless $\bar{a} = \bar{o}$ or $\bar{b} = \bar{o}$, the projection of $\bar{a}$ on $\bar{b}$ vanishes. Consequently, unless $\bar{a} = \bar{o}$ or $\bar{b} = \bar{o}$, the plane through the endpoint of $\bar{a}$ perpendicular to the line through $\bar{b}$ passes through O and hence contains $\bar{a}$. It follows that the line through $\bar{a}$ is perpendicular to the line through $\bar{b}$. It is convenient to say that two vectors $\bar{a}$ and $\bar{b}$ are perpendicular (to each other) if the lines through $\bar{a}$ and $\bar{b}$ are perpendicular, or if either $\bar{a} = \bar{o}$ or $\bar{b} = \bar{o}$. We shall use the symbol $\bar{a} \perp \bar{b}$ to indicate this condition. Our result can then be expressed simply by saying

$$\bar{a} \cdot \bar{b} = 0 \quad \text{implies} \quad \bar{a} \perp \bar{b}.$$

The converse,

$$\bar{a} \perp \bar{b} \quad \text{implies} \quad \bar{a} \cdot \bar{b} = 0,$$

could easily be derived by related arguments.

We are now ready to prove geometric theorems by means of vector operations. We shall give two examples.

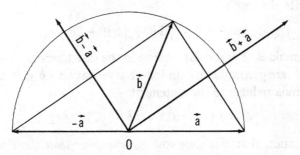

Figure 17. Theorem of Thales expressed in terms of vectors

As our first example we shall prove the theorem of Thales, which states that the two segments connecting the two endpoints of the diameter of a circle with a third point on the periphery are perpendicular. Let us denote by $\vec{a}$ the vector which leads from the center of the circle, taken as initial point O, to one endpoint of the diameter (see Figure 17); then the vector leading to the other endpoint is $-\vec{a}$. Let $\vec{b}$ be the vector leading from O to the third point on the periphery, and consider the two vectors leading to this third point from the endpoints of the diameter. Clearly, when these vectors are transplanted to the origin as initial point, they go over into the vectors $\vec{b} - \vec{a}$ and $\vec{b} - (-\vec{a}) = \vec{b} + \vec{a}$, respectively. To prove that these two vectors are perpendicular we need only show that their inner product vanishes. From rules III$_2'$ and V$_2$, we have

$$(\vec{b} + \vec{a}) \cdot (\vec{b} - \vec{a}) = (\vec{b} + \vec{a}) \cdot \vec{b} - (\vec{b} + \vec{a}) \cdot \vec{a};$$

from rule III$_2$ we conclude further that

$$(\vec{b} + \vec{a}) \cdot (\vec{b} - \vec{a}) = \vec{b} \cdot \vec{b} + \vec{a} \cdot \vec{b} - \vec{b} \cdot \vec{a} - \vec{a} \cdot \vec{a} = |\vec{b}|^2 - |\vec{a}|^2,$$

using the formula for the absolute value and the fact that $\vec{b} \cdot \vec{a} = \vec{a} \cdot \vec{b}$ by rule IV$_2$. Now, since the endpoint of $\vec{a}$ and the endpoint of $\vec{b}$ lie on the same circle with O as center, we have $|\vec{b}| = |\vec{a}|$ and so $(\vec{b} + \vec{a}) \cdot (\vec{b} - \vec{a}) = 0$. Thus we have proved the theorem of Thales simply by relying on the formal rules of vector operations.

To lead up to our second example we employ our formal rules to evaluate the square of the sum $\vec{a} + \vec{b}$ of two vectors $\vec{a}$ and $\vec{b}$. From our rules III$_2$ we obtain

$$(\vec{a} + \vec{b}) \cdot (\vec{a} + \vec{b}) = \vec{a} \cdot \vec{a} + \vec{b} \cdot \vec{a} + \vec{a} \cdot \vec{b} + \vec{b} \cdot \vec{b}.$$

Using the formula for the absolute value, together with rule IV_2, we may write the result as

$$|\vec{a} + \vec{b}|^2 = |\vec{a}|^2 + |\vec{b}|^2 + 2\vec{a} \cdot \vec{b}.$$

This formula is of particular interest in the case where the two vectors $\vec{a}$ and $\vec{b}$ are perpendicular. In this case we have $\vec{a} \cdot \vec{b} = 0$, and hence the formula reduces to the statement

$$|\vec{a} + \vec{b}|^2 = |\vec{a}|^2 + |\vec{b}|^2 \quad \text{if} \quad \vec{a} \perp \vec{b}.$$

We maintain that *this statement is none other than the Pythagorean theorem*. To see this we need only interpret the statement geometrically; see Figure 18. Evidently, the arrows $\vec{a}$, $\vec{a} + \vec{b} = \vec{c}$ and the transplanted arrow $(\vec{b})$ form a right triangle whose sides have the lengths $|\vec{a}|$, $|\vec{a} + \vec{b}| = |\vec{c}|$, and $|\vec{b}|$. Denoting these lengths respectively by a, c, and b, we see that the above formula indeed reduces to the formula $c^2 = a^2 + b^2$.

Thus we have attained our second major aim: the derivation of the Pythagorean theorem simply by applying formal rules of operation to appropriately defined new entities.

One may ask whether or not the present proof of the Pythagorean theorem is "shorter" than the one described in Chapter 1 or any of the other elementary proofs. The answer depends, of course, on how long and how short one considers the various steps that were gone through in a complete proof. If one were to spell out explicitly all the steps needed to derive the formal rules from the facts* of elementary geometry, one would come to regard the present proof as "long"**. But that should not be considered as a shortcoming.

While all the "elementary" proofs of the Pythagorean theorem require specific—more or less ingenious—inventions, no such specific

---

* It is interesting to note that the law of proportionality was essentially used in deriving the multiplication laws III and IV for vectors, while the proofs given in Chapter 1 used only the laws of parallel lines and the laws of congruences.

** However, a "short" proof can be extracted from it: Let $\vec{a}_c$ and $(\vec{b})_c$ be the projections of $\vec{a}$ and $(\vec{b})$ on $\vec{c}$. Flipping the triangles spanned by $\vec{a}$, $\vec{a}_c$ and $(\vec{b})$, $(\vec{b})_c$, where $(\vec{b})_c$ is attached to the endpoint of $\vec{a}_c$, and applying the same argument that was used for proving rule IV_2, one finds the proportions $|\vec{a}_c|/|\vec{a}| = |\vec{a}|/|\vec{c}|$ and $|\vec{b}_c|/|\vec{b}| = |\vec{b}|/|\vec{c}|$ or $|\vec{c}||\vec{a}_c| = |\vec{a}|^2$, $|\vec{c}||\vec{b}_c| = |\vec{b}|^2$, whence $|\vec{c}|^2 = |\vec{a}|^2 + |\vec{b}|^2$ follows by addition.

ingenuity was needed for the present proof. It just fell into our lap. Of course, invention and discovery were involved too—the invention of the concept of vector, and the discovery that operations can be so defined that simple formal rules are valid. But here the invention and discovery are unspecific; they are of quite general significance and form the broad basis on which large units of mathematics and mathematical physics are built.

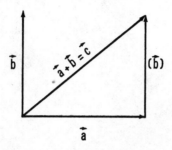

Figure 18. Pythagorean theorem expressed in terms of vectors

CHAPTER FOUR

Components and Coordinates. Spaces of Higher Dimension

As we have seen in Chapter 3, it is possible to introduce operations with vectors having formal properties very similar to the formal properties which characterize the ordinary algebraic operations with numbers. One may wonder whether or not one can actually reduce vector operations to ordinary algebraic operations. This is indeed possible. In fact, the reduction of geometrical operations to operations with numbers, initiated by Descartes, has proved to be a decisive step—if not the most decisive one—in the development of geometry. This general reduction has made it possible to reduce all specific computational work in geometry to computations with numbers and to derive geometrical theorems from theorems on operations with numbers. Moreover, this reduction has made it easy to free geometry from the restriction to our three dimensional space and to extend the concepts of geometry to so-called spaces of more than three dimensions, in fact to spaces of infinitely many dimensions. In all this work vectors are the most effective tool.

The notion of vector and the rules of operations with vectors were developed during the second half of the nineteenth century in a rather roundabout way. The proper understanding of vectors and vector operations was attained about 1880 through the work of J. Willard Gibbs of Yale.

In the following discussion—just for the sake of convenience—we shall restrict ourselves to the plane and assume all vectors to lie in this plane. All that we are going to say in the main part of this chapter can immediately be carried over to vectors in three dimensional space; this we shall discuss at the end of the chapter.

Our first aim is to characterize every vector by two numbers, its "components". (A three dimensional vector would be characterized by three components.) To this end we introduce a system of "unit vectors" and represent every vector as a "linear combination" of them.

A unit vector $\vec{u}$ is a vector of magnitude 1, i.e. one with $|\vec{u}| = 1$. A "system of orthogonal unit vectors" in our plane is a pair of unit vectors which are perpendicular to each other. Distinguishing the two vectors by "superscripts" (1) and (2), we may write the relations that characterize such a pair as

(*) $|\vec{u}^{(1)}| = 1,$ $|\vec{u}^{(2)}| = 1,$ $\vec{u}^{(1)} \cdot \vec{u}^{(2)} = 0.$

(Instead of the term "perpendicular" one more frequently uses the term "orthogonal".)

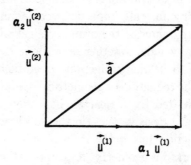

Figure 19. Components of a vector with respect to a pair of orthogonal unit vectors

In order to select such a system of unit vectors one may first select any straight line and then take as $\vec{u}^{(1)}$ one of the two unit vectors in the direction of this line. As $\vec{u}^{(2)}$ one then may choose one of the two unit vectors perpendicular to $\vec{u}^{(1)}$. We shall take as $\vec{u}^{(2)}$ the unit vector that can be obtained from the vector $\vec{u}^{(1)}$ by rotating it through a right angle in the counter clockwise direction; see Figure 19.

It is an important fact that every vector $\vec{a}$ in the plane can be written as a "linear combination"

$$\vec{a} = \alpha_1 \vec{u}^{(1)} + \alpha_2 \vec{u}^{(2)}$$

of the two unit vectors $\bar{u}^{(1)}$, $\bar{u}^{(2)}$ with two appropriately chosen real numbers α_1, α_2.

Before proving this we first suppose that a vector $\bar{a}$ is known to be such a linear combination. Then, we claim, the two numbers α_1, α_2 can simply be expressed as the inner products of the vector $\bar{a}$ and the two unit vectors. To show this, one need only multiply the above expression for $\bar{a}$ on the left by $\bar{u}^{(1)}$ and $\bar{u}^{(2)}$; using the basic relations (*) one finds the desired relations

$$\bar{u}^{(1)} \cdot \bar{a} = \alpha_1, \qquad \bar{u}^{(2)} \cdot \bar{a} = \alpha_2.$$

From this result we shall conclude that the vector $\bar{a}$ cannot be written as a linear combination of $\bar{u}^{(1)}$, $\bar{u}^{(2)}$ with numbers β_1, β_2 different from α_1, α_2. In other words, the two numbers α_1, α_2 are uniquely determined. For, suppose we could write $\bar{a}$ in the form $\bar{a} = \beta_1\bar{u}^{(1)} + \beta_2\bar{u}^{(2)}$ with the pair of numbers β_1, β_2 which might be different from the pair α_1, α_2; then we could again multiply this linear combination by $\bar{u}^{(1)}$ and $\bar{u}^{(2)}$. The result

$$\bar{u}^{(1)} \cdot \bar{a} = \beta_1, \quad \bar{u}^{(2)} \cdot \bar{a} = \beta_2 \qquad \text{would lead to} \qquad \beta_1 = \alpha_1, \quad \beta_2 = \alpha_2$$

and thus imply that indeed α_1 and α_2 are unique.

While so far we have assumed of the vector $\bar{a}$ that it is a linear combination of $\bar{u}^{(1)}$ and $\bar{u}^{(2)}$, we now drop this assumption and prove that any vector $\bar{a}$ in the plane is such a combination. Moreover, defining numbers α_1, α_2 by the formulas $\alpha_1 = \bar{u}^{(1)} \cdot \bar{a}$, $\alpha_2 = \bar{u}^{(2)} \cdot \bar{a}$, we shall prove that $\bar{a}$ is exactly the combination $\bar{a} = \alpha_1\bar{u}^{(1)} + \alpha_2\bar{u}^{(2)}$. To this end we need only show that the vector

$$\bar{v} = \bar{a} - \alpha_1\bar{u}^{(1)} + \alpha_2\bar{u}^{(2)}$$

is zero. Now, the inner product of this vector $\bar{v}$ with $\bar{u}^{(1)}$ is

$$\bar{u}^{(1)} \cdot \bar{v} = \bar{u}^{(1)} \cdot \bar{a} - \alpha_1\bar{u}^{(1)} \cdot \bar{u}^{(1)} - \alpha_2\bar{u}^{(1)} \cdot \bar{u}^{(2)} = \bar{u}^{(1)} \cdot \bar{a} - \alpha_1 = 0.$$

Hence $\bar{v}$ is perpendicular to $\bar{u}^{(1)}$ and therefore has the direction of $\bar{u}^{(2)}$. Similarly, one shows that $\bar{v}$ has just as well the direction of $\bar{u}^{(1)}$. Obviously, the only vector having the direction of $\bar{u}^{(1)}$ as well as that of $\bar{u}^{(2)}$ is the zero vector. Thus we conclude $\bar{v} = \bar{o}$, and our statement is established.

This result may be formulated by saying that a vector $\bar{a}$ is completely characterized by the two numbers α_1, α_2, called its "components" with respect to the unit system $\{\bar{u}^{(1)}, \bar{u}^{(2)}\}$. We express this fact also by writing

$$\bar{a} = (\alpha_1, \alpha_2).$$

We proceed to show how the *vector operations* express themselves in terms of components. Let $\bar{a} = (\alpha_1, \alpha_2)$ and $b = (\beta_1, \beta_2)$ be two vectors. Then, as is immediately verified from rule III$_1'$, the sum of these vectors has the components $\alpha_1 + \beta_1$ and $\alpha_2 + \beta_2$ so that it may be described by

$$\bar{a} + \bar{b} = (\alpha_1 + \beta_1, \ \alpha_2 + \beta_2).$$

Thus we see that *addition of vectors* is reduced to *addition of their components*.

Similarly, multiplication of a vector $\bar{a} = (\alpha_1, \alpha_2)$ by a number λ leads to the vector

$$\lambda \, \bar{a} = (\lambda\alpha_1, \lambda\alpha_2),$$

as is readily verified from rules III$_1$ and V$_1'$. Thus *multiplication of a vector by a number* is reduced to *multiplication of its components by this number*.

Less obvious is the expression of the inner product $\bar{a} \cdot \bar{b}$ of two vectors $\bar{a}$ and $\bar{b}$ in terms of their components. To derive this expression we write

$$\bar{a} = \alpha_1\bar{u}^{(1)} + \alpha_2\bar{u}^{(2)}, \qquad \bar{b} = \beta_1\bar{u}^{(1)} + \beta_2\bar{u}^{(2)}$$

and then obtain

$$\bar{a} \cdot \bar{b} = (\alpha_1\bar{u}^{(1)} + \alpha_2\bar{u}^{(2)}) \cdot (\beta_1\bar{u}^{(1)} + \beta_2\bar{u}^{(2)})$$

$$= \alpha_1\beta_1\bar{u}^{(1)} \cdot \bar{u}^{(1)} + \alpha_1\beta_2\bar{u}^{(1)} \cdot \bar{u}^{(2)} + \alpha_2\beta_1\bar{u}^{(2)} \cdot \bar{u}^{(1)} + \alpha_2\beta_2\bar{u}^{(2)} \cdot \bar{u}^{(2)}.$$

Using identities (*) we arrive at the fundamental formula

$$\bar{a} \cdot \bar{b} = \alpha_1\beta_1 + \alpha_2\beta_2.$$

Observe that this formula expresses *the inner product of two vectors as the sum of the products of their components*.

An important special case of this formula arises in case $\bar{b} = \bar{a}$. In this case we have $\bar{a} \cdot \bar{a} = |\bar{a}|^2$ and our formula for the inner product yields

$$|\bar{a}|^2 = \alpha_1^2 + \alpha_2^2 \qquad \text{and} \qquad |\bar{a}| = \sqrt{\alpha_1^2 + \alpha_2^2}.$$

Clearly, the first formula here may be regarded as an expression of the Pythagorean theorem; for, the vector $\bar{a}$ runs along the hypotenuse of the right triangle formed by the vector $\alpha_1\bar{u}^{(1)}$ and the transplanted vector $\alpha_2\bar{u}^{(2)}$, while the absolute values $|\alpha_1|$, $|\alpha_2|$ of the components α_1, α_2 are evidently the lengths of these legs. This new form of the Pythagorean theorem will later on be used as the starting point for the development of geometry in a space of more than three dimensions.

Now that vectors have been completely described by pairs of numbers, that is, by their components with respect to a pair of orthogonal unit vectors, it is evident that *one can deduce the formal rules for addition and multiplication of vectors from the corresponding formal rules for numbers.* We shall not carry this out; we only mention as an example that the validity of rule IV$_2$ follows immediately from the fact that the inner product of $\bar{a}$ and $\bar{b}$ is given by $\alpha_1\beta_1 + \alpha_2\beta_2$, since, for ordinary real numbers, $\alpha_1\beta_1 = \beta_1\alpha_1$ and $\alpha_2\beta_2 = \beta_2\alpha_2$. In any case it is clear that all these matters become so much simpler than they were before as soon as one considers vectors as given by two numbers. This fact naturally leads to a *change of attitude*. Instead of beginning geometry with the Euclidean axioms and postulates, one may simply begin geometry by introducing points—and vectors leading to them— as entities which are characterized by two numbers α_1, α_2. Addition of the vectors $\bar{a} = (\alpha_1, \alpha_2)$ and $\bar{b} = (\beta_1, \beta_2)$ is then *defined* by $\bar{a} + \bar{b} = (\alpha_1 + \beta_1, \alpha_2 + \beta_2)$, and multiplication of a vector $\bar{a}$ by a number λ is then *defined* by $\lambda\bar{a} = (\lambda\alpha_1, \lambda\alpha_2)$. Finally, inner multiplication of two vectors is *defined* by $\bar{a} \cdot \bar{b} = \alpha_1\beta_1 + \alpha_2\beta_2$, and the magnitude $|\bar{a}|$ of a vector is *defined* by $|\bar{a}|^2 = \alpha_1^2 + \alpha_2^2$. The validity of the formal rules can then be immediately verified.

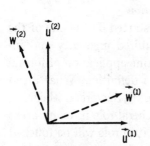

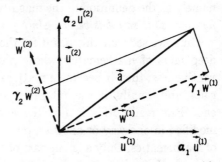

Figure 20. Rotation of a system of orthogonal unit vectors

Figure 21. Components of a vector with respect to two systems of unit vectors

In this approach the two unit vectors $\bar{u}^{(1)} = (1, 0)$ and $\bar{u}^{(2)} = (0, 1)$ appear to play a special role; but actually this special role is irrelevant. For, we may just as well choose any other pair of unit vectors $\bar{w}^{(1)}$ and $\bar{w}^{(2)}$ (given by their components with respect to $\bar{u}^{(1)}$ and $\bar{u}^{(2)}$) and express all vectors as linear combinations $\bar{a} = \gamma_1 \bar{w}^{(1)} + \gamma_2 \bar{w}^{(2)}$ of these new unit vectors; see Figures 20, 21. Addition of vectors and multiplication of vectors by constants is then expressed in terms of the new components γ_1, γ_2 in the same manner as it was expressed in terms of the original components α_1, α_2.

Moreover, if the new unit vectors are orthogonal, i.e., if they satisfy the relations

$$(**) \qquad \bar{w}^{(1)} \cdot \bar{w}^{(1)} = \bar{w}^{(2)} \cdot \bar{w}^{(2)} = 1, \qquad \bar{w}^{(1)} \cdot \bar{w}^{(2)} = 0$$

(as we have assumed in Figures 20, 21), the inner product of two vectors is expressed in terms of the new components in the same manner as it was expressed in terms of the original ones. (This can be immediately verified.) For the inner product of a vector $\bar{a}$ with itself, in particular, this statement yields the formula

$$|\bar{a}|^2 = \gamma_1^2 + \gamma_2^2.$$

Note that in the present approach the magnitude $|\bar{a}|$ of the vector $\bar{a} = (\alpha_1, \alpha_2)$—or the distance of its endpoint from the origin O—is *defined* by the formula $|\bar{a}|^2 = \alpha_1^2 + \alpha_2^2$ with respect to the original coordinate system, and that the way in which the new unit vectors were introduced entails that the square of the magnitude of the vector $\bar{a}$ is given by $|\bar{a}|^2 = \gamma_1^2 + \gamma_2^2$. Obviously, these formulas correspond to the Pythagorean theorem. In particular, inasmuch as the first of these formulas for $|\bar{a}|^2$ simply served as a definition, namely as the definition of the magnitude of a vector, *the Pythagorean formula, so to say, has become simply a definition.*

Thus, we see that this beautiful theorem started out as one of the deep facts of mathematics whose proof required ingenuity and inventiveness; then it became a matter of routine application of formal rules, and, finally, it was relegated to just a definition. What a sad end! But really, this is not the end. The Pythagorean theorem, modified in one way or another, plays the central role in many areas of mathematics. Only a small part of this central role will be touched on in the following discussion.

We shall now show that a geometry of any number of dimensions can be developed very easily if one adopts the new attitude described above and accepts the Pythagorean theorem as a definition*.

A point in n-dimensional space is defined as an entity which can be described by n coordinates $\alpha_1, \alpha_2, \cdots, \alpha_n$. These coordinates may also be regarded as the components of the vector leading to this point from the origin $O = (0, 0, \cdots, 0)$. The inner product $\bar{a} \cdot \bar{b}$ of two vectors $\bar{a} = (\alpha_1, \alpha_2, \cdots, \alpha_n)$ and $\bar{b} = (\beta_1, \beta_2, \cdots, \beta_n)$ is defined by the formula

$$\bar{a} \cdot \bar{b} = \alpha_1\beta_1 + \alpha_1\beta_1 + \cdots + \alpha_n\beta_n.$$

The magnitude of the vector $\bar{a}$ is defined as

$$|\bar{a}| = \sqrt{\alpha_1^2 + \alpha_2^2 + \cdots + \alpha_n^2};$$

the formula for its square is

$$|\bar{a}|^2 = \alpha_1^2 + \alpha_2^2 + \cdots + \alpha_n^2$$

and may be regarded as the n-dimensional analogue of the Pythagorean theorem.

Two vectors $\bar{a}$ and $\bar{b}$ are now defined to be orthogonal if $\bar{a} \cdot \bar{b} = 0$. The n unit vectors

$$\bar{u}^{(1)} = (1, 0, \cdots, 0), \quad \bar{u}^{(2)} = (0, 1, \cdots, 0), \quad \cdots, \quad \bar{u}^{(n)} = (0, \cdots, 0, 1)$$

evidently form an orthogonal system of unit vectors, that is to say, a system of unit vectors having the property that any two different ones are orthogonal.

Other orthogonal systems $\{\hat{u}^{(1)}, \hat{u}^{(2)}, \cdots, \hat{u}^{(n)}\}$ of unit vectors may be introduced. (Here we have omitted the qualifier →.) The components of any vector $\bar{a}$ with respect to such a system are defined to be the numbers

$$\hat{\alpha}_1 = \hat{u}^{(1)} \cdot \bar{a}, \quad \hat{\alpha}_2 = \hat{u}^{(2)} \cdot \bar{a}, \quad \cdots, \quad \hat{\alpha}_n = \hat{u}^{(n)} \cdot \bar{a}.$$

We then find, by arguments similar to those used for the plane, that the vector $\bar{a}$ can be written as

$$\bar{a} = \hat{\alpha}_1\hat{u}^{(1)} + \hat{\alpha}_2\hat{u}^{(2)} + \cdots + \hat{\alpha}_n\hat{u}^{(n)}.$$

* There is still another powerful approach to geometry of any number of dimensions: vectors are introduced from the beginning as entities for which operations obeying the rules of Chapter 3 can be defined. However, the approach described here seems to be more suitable as background for the material discussed in Chapter 7.

For the inner product of two vectors $\vec{a}$ and $\vec{b}$ in terms of these new components, we obtain the expression

$$\vec{a} \cdot \vec{b} = \hat{\alpha}_1\hat{\beta}_1 + \hat{\alpha}_2\hat{\beta}_2 + \cdots + \hat{\alpha}_n\hat{\beta}_n$$

simply by using the orthogonality of the unit vectors $\hat{u}^{(1)}$, $\hat{u}^{(2)}$, $\cdots$, $\hat{u}^{(n)}$.

All basic notions of two or three dimensional geometry can be carried over to n dimensions by analogy. For example, geometrical entities such as straight lines or planes may easily be introduced. A straight line through the point O in the direction of the vector $\vec{a}$ is formed by the endpoints of the multiples $\lambda\vec{a}$ of the vector $\vec{a}$; here λ stands for any real number, positive, negative, or zero. A plane through O may be described as consisting of all the endpoints of the vectors

$$\lambda\vec{a} + \mu\vec{b}$$

where $\vec{a}$, $\vec{b}$ are two given vectors not parallel to each other, while λ and μ range over all real numbers.

Similarly, one may define what is called a "hyperplane" of any dimension k less than n as the set of endpoints of all vectors of the form

$$\lambda_1\vec{a}^{(1)} + \lambda_2\vec{a}^{(2)} + \cdots + \lambda_k\vec{a}^{(k)}.$$

Here λ_1, λ_2, $\cdots$, λ_k are any real numbers, and the k vectors $\vec{a}^{(1)}$, $\vec{a}^{(2)}$, $\cdots$, $\vec{a}^{(k)}$ are such that no linear combination of them is the zero vector unless all numbers λ_1, λ_2, $\cdots$, λ_k are zero.

A "hypersphere" with center O and radius R is defined as the set of all endpoints of vectors $\vec{a}$ of length R.

These remarks may perhaps suffice to indicate how easily the notions of Euclidean geometry can be carried over to n-dimensional geometry.

One may wonder whether or not it is possible to go a big step further and introduce an *infinite-dimensional* space of vectors

$$\vec{a} = (\alpha_1, \alpha_2, \cdots)$$

with infinitely many components, defining the magnitude $|\vec{a}|$ by an infinite-dimensional analogue of the Pythagorean formula, i.e. by the infinite series

$$|\vec{a}|^2 = \alpha_1^2 + \alpha_2^2 + \cdots .$$

In fact, this can be done; the resulting "Hilbert space" and its properties have been studied extensively since the beginning of this century. As should be expected, new features appear in the theory of this space, connected with the question of convergence of infinite series, a question which has no counterpart in finite-dimensional geometry. We refrain from any discussion of these matters.

CHAPTER FIVE

Momentum and Energy. Elastic Impact

The notion of vector in three dimensional space is one of the important concepts used in the mathematical description of physical entities. The velocity of a piece of matter has a direction and a magnitude; it is a vector. The force that is exerted at a point of a body is also a vector. In this chapter we shall be concerned only with two kinds of vectors, the "velocity vector" and the closely related "momentum vector".

Different portions of a piece of matter—referred to as a "body"—may have different velocities. We shall assume that this is not so for the bodies to be considered in the present chapter; that is to say, we shall assume that all parts of the body move with the same velocity. When this is the case, the body is frequently imagined to be concentrated at a single mathematical point (at its center of gravity, for example) and is called a "particle". This single point serves as initial point of the velocity vector.

Before numerical quantities can be assigned to the motion of a body, a unit of length and a unit of time must be selected; we assume that this has been done.

Suppose now a body moves in a definite direction. During an interval of time it will cover a certain distance. An interval of time is a fraction (or multiple) of the unit time; when we speak of the interval τ of time, where τ is a positive number, we mean τ times the unit interval of time. Similarly, when we speak of the distance δ, we mean δ times the unit length.

Suppose now the body moves—in an unchanging direction—in such a way that the ratio δ/τ, where δ is the distance it covers in a time interval τ, is the same no matter how small (or large) the interval τ is taken. Then we say the body moves with constant velocity. The ratio δ/τ is called the "speed" of the body, while the velocity of the body is the vector $\vec{v}$ whose direction is that of the motion of the body and whose magnitude is its speed. Since the body is assumed to move with constant velocity, this vector $\vec{v}$ is transplanted parallel to itself during the motion.

The notions of addition and multiplication of velocity vectors play an important role in the description of motion.

Consider a particle that moves with the constant velocity $\vec{v}$ and suppose its motion is observed from another body, called the "platform", which itself moves with a constant velocity, $\vec{v}_1$ say. Let $\vec{v}_2$ be the velocity of the particle as observed from the moving platform. Then, we state, the velocity $\vec{v}$ of the particle is given by

$$\vec{v} = \vec{v}_1 + \vec{v}_2.$$

We can imagine that the "original" observation that the body moved with velocity $\vec{v}$ was made from a resting platform, but we can also think of the original observation platform as moving with some constant velocity; in either case the velocity $\vec{v}$ of the particle with respect to the original platform is $\vec{v}_1 + \vec{v}_2$.

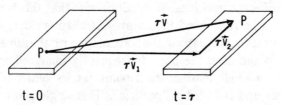

Figure 22. Addition of velocities

To justify this statement we note that the point on the platform moving with velocity $\vec{v}_1$, which was at the initial point O at some initial time, will be at the endpoint of the vector $\tau\vec{v}_1$ after the time interval τ; see Figure 22. The moving particle will at this time be at the endpoint of the vector $\tau\vec{v}_2$ when this vector is attached to the endpoint of $\tau\vec{v}_1$, that is to say, it will be at the endpoint of $\tau\vec{v}_1 + \tau\vec{v}_2$

according to the definition of vector addition. On the other hand, it is clear that, after the time τ, the particle will be at the endpoint of the vector $\tau\bar{v}$. Consequently the relation $\tau\bar{v} = \tau\bar{v}_1 + \tau\bar{v}_2$ obtains and hence the statement follows.

The expression for the speed $|\bar{v}|$ of a particle with velocity $\bar{v}$ is obtained from the inner product of $\bar{v}$ with itself:

$$|\bar{v}|^2 = \bar{v} \cdot \bar{v}.$$

Setting $\bar{v} = \bar{v}_1 + \bar{v}_2$, with $\bar{v}_1$ being the velocity of a platform as above, we find

$$|\bar{v}|^2 = |\bar{v}_1|^2 + 2\bar{v}_1 \cdot \bar{v}_2 + |\bar{v}_2|^2.$$

In particular, if platform and particle move in perpendicular directions, we have

$$|\bar{v}|^2 = |\bar{v}_1|^2 + |\bar{v}_2|^2,$$

a formula related to the Pythagorean theorem.

If a particle is under the influence of forces its velocity changes in the course of time. To describe the motion of such a particle, it is necessary to know how these forces act and how they affect its velocity. We shall not discuss these matters. We shall confine ourselves to discussing a particular type of motion—the fundamental process called "impact"—that can be determined in essential features without reference to these forces. Impact occurs if two particles move toward each other, hit each other, and bounce back from each other.

Suppose we know the velocities $\bar{v}_1$, $\bar{v}_2$ of the two particles before impact; what are their velocities afterwards? The answer to this question can be derived from two basic principles of mechanics: the *law of conservation of momentum* and the *law of conservation of energy*.

To formulate these laws it is necessary to introduce the notion of mass, a number (measured in an appropriate unit) assigned to each particle. We do not intend to give a complete characterization of this quantity, but we should like to describe the main steps of one possible characterization.

First one selects a particular piece of matter and assigns the mass 1 to it. One then regards any other piece of the same material as congruent to a multiple or fraction of the unit piece and assigns to this other piece the same multiple or fraction of unity as its mass.

To assign masses to pieces of different material, it is sufficient to stipulate when two of them have the same mass. The customary stipulation is that they should balance on symmetrical scales; in other words, they should have the same weight. Clearly, reference to the force of gravity is implied in this stipulation. We wish to avoid reference to forces in the present context. This can be done by employing a symmetrical impact process, imagining that the two pieces move with the same speed towards each other on a straight line and then bounce back without undergoing any change. If in moving away from each other they again have the same speed, we shall say that they have the same mass. We add that the mass of a "compound particle", consisting of just two particles of any material staying together, is simply the sum of the masses of the two "component" particles.

We now can define the *momentum* of a particle with mass m and velocity $\bar{v}$ as the vector

$$\bar{p} = m\bar{v}.$$

The *kinetic energy* of such a particle is defined as the number

$$e = \tfrac{1}{2}m \mid \bar{v} \mid^2$$

and is the only kind of energy that will be referred to in this chapter; other kinds of energy will be referred to in Chapter 6.

The first law that is obeyed in any impact process says that the "total momentum"

$$\bar{p}_1 + \bar{p}_2 = m_1\bar{v}_1 + m_2\bar{v}_2$$

of the two particles—with masses m_1, m_2 and velocities $\bar{v}_1$, $\bar{v}_2$—is the same before and after impact. The importance of the notion of momentum stems to a large degree from its role in the first impact law.

This "momentum law" implies a relation between the velocities of the particles after and before impact. To describe this relation, we denote for a moment by $[\bar{v}_1]$ and $[\bar{v}_2]$ the differences of the velocity vectors of the two particles before and after impact and write the momentum law in the form $m_1[\bar{v}_1] + m_2[\bar{v}_2] = \bar{o}$, or in the equivalent form

$$m_1[\bar{v}_1] = -m_2[\bar{v}_2].$$

We then see that, according to this law, the changes of velocity of the two particles are inversely proportional to their masses and opposite in direction. To determine the magnitude of these changes a second law is needed; this will be the energy law to be formulated a little later.

The momentum law holds for all impact processes, as could be derived from the principles of Newton's mechanics. The main ingredient in such a derivation is the law that the force exerted by one particle on the other during impact is opposite in direction and equal in magnitude to the force exerted by the second particle on the first one. The proper form of the energy law, on the other hand, depends on the specific type of impact process considered, which in turn depends on the nature of the colliding particles and of the forces acting between them.

In the present chapter we shall consider only one type of impact, "elastic" impact. An essential feature of elastic impact is that each particle retains all of its physical characteristics except its velocity during the process; just the velocities change. In an actual piece of matter this feature can be realized only approximately, in general in good approximation only if this piece of matter may be regarded as a particle concentrated at a single point in space, and if the process of impact may be regarded as taking place at approximately a single instant, and not during a time interval. Since we do not want to discuss those parts of mechanics that deal with forces, we shall not give a precise characterization of an impact process as "elastic" in terms of the forces acting between the particles. As a result of such a characterization one could derive from Newton's mechanics that in an elastic impact the "total kinetic energy"

$$e_1 + e_2 = \tfrac{1}{2}m_1 \,|\, \bar{v}_1 \,|^2 + \tfrac{1}{2}m_2 \,|\, \bar{v}_2 \,|^2$$

of the two particles is the same before and after impact. In the following we shall simply assume that for elastic impact the momentum and energy laws hold as stated.

The concepts of energy and momentum pervade all of physics. No justice can be done to them by a brief and abstract explanation of their meaning. Still, the subsequent discussion of impact throws some light on the basic role that these notions play in the mathematical description of nature.

We proceed to discuss the two conservation laws somewhat more closely. We shall characterize vectors before impact by a hat $\wedge$ and those afterwards by an inverted hat $\vee$ and omit the qualifier $\rightarrow$.

We may read* these new qualifiers $\wedge$ and $\vee$ as "in" and "out". Thus the incoming and outgoing momenta are written as

$$\hat{p}_1 = m_1\hat{v}_1, \quad \hat{p}_2 = m_2\hat{v}_2, \quad \text{and} \quad \check{p}_1 = m_1\check{v}_1, \quad \check{p}_2 = m_2\check{v}_2.$$

The two conservation laws can now be written in the form

$$m_1\check{v}_1 + m_2\check{v}_2 = m_1\hat{v}_1 + m_2\hat{v}_2,$$

$$\tfrac{1}{2}m_1 | \check{v}_1 |^2 + \tfrac{1}{2}m_2 | \check{v}_2 |^2 = \tfrac{1}{2}m_1 | \hat{v}_1 |^2 + \tfrac{1}{2}m_2 | \hat{v}_2 |^2.$$

With the aid of these equations we should like to determine the state after impact from the known state before impact. Before explaining how to do this we bring up a question of general significance concerning these two laws.

Suppose a particle, moving with the velocity $\bar{v}$ with respect to a resting platform, is observed from an observation platform which itself moves with the velocity $\bar{v}_0$. If the observer does not know, or disregards, that the platform moves, he will ascribe the velocity $\bar{v} - \bar{v}_0$ to the particle. Consequently, he will ascribe the momentum $m(\bar{v} - \bar{v}_0)$ and the energy $\tfrac{1}{2}m | \bar{v} - \bar{v}_0 |^2$ to it. Naturally, one will ask: do the conservation laws also hold with respect to these "relative" energies and momenta? They do!

With $\hat{v}_1 - \bar{v}_0$, $\hat{v}_2 - \bar{v}_0$ and $\check{v}_1 - v_0$, $\check{v}_2 - v_0$ in place of $\hat{v}_1$, $\hat{v}_2$ and $\check{v}_1$, $\check{v}_2$, the momentum law takes the form

$$m_1\check{v}_1 + m_2\check{v}_2 - m_1\bar{v}_0 - m_2\bar{v}_0 = m_1\hat{v}_1 + m_2\hat{v}_2 - m_1\bar{v}_0 - m_2\bar{v}_0;$$

the terms involving $\bar{v}_0$ are the same on both sides, while the others cancel each other by virtue of the original momentum law, that is, the one with respect to the resting platform.

The energy law takes the form

$$
\left.
\begin{aligned}
&\tfrac{1}{2}m_1 | \check{v}_1 |^2 + \tfrac{1}{2}m_2 | \check{v}_2 |^2 \\[4pt]
&- (m_1\bar{v}_0 \cdot \check{v}_1 + m_2\bar{v}_0 \cdot \check{v}_2) \\[4pt]
&+ \tfrac{1}{2}m_1 | \bar{v}_0 |^2 + \tfrac{1}{2}m_2 | \bar{v}_0 |^2
\end{aligned}
\right\}
=
\left\{
\begin{aligned}
&\tfrac{1}{2}m_1 | \hat{v}_1 |^2 + \tfrac{1}{2}m_2 | \hat{v}_2 |^2 \\[4pt]
&- (m_1\bar{v}_0 \cdot \hat{v}_1 + m_2\bar{v}_0 \cdot \hat{v}_2) \\[4pt]
&+ \tfrac{1}{2}m_1 | \bar{v}_0 |^2 + \tfrac{1}{2}m_2 | \bar{v}_0 |^2.
\end{aligned}
\right.
$$

The terms on both sides of the third line are the same. The terms in the first line cancel each other by virtue of the original energy law.

---

* Visualizing time as running from the bottom to the top of the page we may readily remember that the hat $\wedge$ refers to two particles which meet, and the inverted hat $\vee$ to two particles which run away from each other.

What about the terms in the middle line? They cancel each other by virtue of the momentum law. To see this one need only take the inner product of both sides of the original momentum law with the vector $\bar{v}_0$.

This consideration shows how closely the notions of energy and momentum are connected. As a matter of fact one could easily derive the momentum law from our energy law, if one assumes that the energy law holds no matter what the velocity of the platform is.

In any case it is a remarkable fact that the notions of energy and momentum depend on the motion of the observer, but the laws of conservation combined do not.

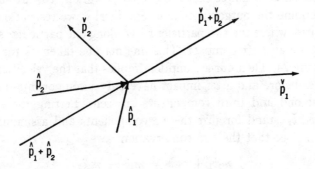

Figure 23. Conservation of total momentum in an impact process

We now take up the problem, mentioned above, of determining what happens after impact when it is known what happens before impact. Note that the momentum equation is a vector equation and actually stands for three equations, one for each of the three components; the energy equation is a single (scalar) equation. Consequently, the two conservation laws represent four equations. The number of unknown quantities, on the other hand, is six: the three components of $\check{v}_1$ and the three components of $\check{v}_2$ (provided the incoming velocities $\hat{v}_1$ and $\hat{v}_2$ are known). See Figure 23.

Thus it is clear that the two conservation laws alone are not sufficient to determine the outgoing velocities when the incoming ones are given. In fact, the motion of the two particles after impact depends on the detailed shape of the particles, on the way they hit each other, and possibly on the manner in which the forces between them act while they are in contact. Our simplification, by which we considered the two particles as points in the strict sense, is too severe. Additional information about what happens during impact is needed.

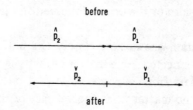

Figure 24. Momenta in a one dimensional impact process

We shall restrict ourselves to a special case in which the additional information is so simple that the two conservation laws are sufficient to determine the outgoing motion. Specifically, we restrict ourselves to the case where the two particles move along one particular straight line before and after impact. This line may be taken as the x-axis; see Figure 24. Then our assumption implies that the velocities of the particles before and after impact have only x-components, i.e., that their second and third components are zero. Letting the symbols $\hat{v}_1$, $\hat{v}_2$, $\check{v}_1$, $\check{v}_2$ stand for just these x-components and assuming $\hat{v}_1$, $\hat{v}_2$ given, we see that the two conservation laws

$$m_1\check{v}_1 + m_2\check{v}_2 = m_1\hat{v}_1 + m_2\hat{v}_2$$

$$\tfrac{1}{2}m_1 \mid \check{v}_1 \mid^2 + \tfrac{1}{2}m_2 \mid \check{v}_2 \mid^2 = \tfrac{1}{2}m_1 \mid \hat{v}_1 \mid^2 + \tfrac{1}{2}m_2 \mid \hat{v}_2 \mid^2$$

represent precisely two equations for the two unknowns $\check{v}_1$, $\check{v}_2$.

Does this pair of equations have a solution? By solution we here mean a pair of numbers $\check{v}_1$, $\check{v}_2$ that satisfies the equations. To facilitate answering this question we may plot the graphs of the two equations. In doing this, it is convenient to take as coordinates the x-components of the two momenta $\bar{p}_1 = m\bar{v}_1$, $\bar{p}_2 = m\bar{v}_2$, rather than the velocities; we denote by $\hat{p}_1$, $\hat{p}_2$ the x-components of the momenta of the two incoming particles, by $\check{p}_1$, $\check{p}_2$ those of the momenta of the two outgoing particles. Then the (x-component of the) total momentum is

$$p_0 = \hat{p}_1 + \hat{p}_2,$$

and the total energy is

$$e_0 = \frac{\hat{p}_1^2}{2m_1} + \frac{\hat{p}_2^2}{2m_2};$$

p_0 and e_0 are assumed to be known inasmuch as $\hat{p}_1$, $\hat{p}_2$ and, of course, m_1, m_2 are known. The outgoing momenta will then be solutions $p_1 = \check{p}_1$, $p_2 = \check{p}_2$ of the equations

$$p_1 + p_2 = p_0, \qquad \frac{p_1^2}{m_1} + \frac{p_2^2}{m_2} = 2e_0.$$

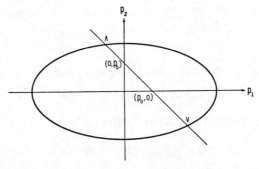

Figure 25. Graph of laws of conservation of momentum and energy for a one dimensional impact process. Graph of energy law is ellipse. Graph of momentum law is straight line

The graph of the first equation is evidently a straight line intersecting the p_1- and p_2-axes at the points $(p_0, 0)$ and $(0, p_0)$; the graph of the second equation is an ellipse with the semi-axes

$$a_1 = \sqrt{2m_1e_0}, \qquad a_2 = \sqrt{2m_2e_0};$$

see Figure 25 and its caption. We are sure that the straight line intersects the ellipse in at least one point; for, one solution of the pair of equations is known to begin with, namely, the incoming pair of momenta $p_1 = \hat{p}_1$, $p_2 = \hat{p}_2$. If the straight line is not tangent to the ellipse there is another solution which one naturally expects to give the outgoing momenta $p_1 = \check{p}_1$, $p_2 = \check{p}_2$.

Before showing that this second solution exists and gives the outgoing momenta we should like to explain how to find it algebraically. If one expresses p_2 as $p_0 - p_1$ from the momentum law and inserts this into the energy law, a quadratic equation for p_1 results. Since one real solution of it is known, the equation can be split into two real linear factors; the other root can then be found by solving a linear equation. Instead of carrying this out, we prefer to proceed a little differently.

In the following it is always assumed that v_1 and v_2 are connected with p_1 and p_2 by

$$v_1 = p_1/m_1, \qquad v_2 = p_2/m_2.$$

We first set $2e_0 = p_1\hat{v}_1 + p_2\hat{v}_2$ and write the energy law in the form

$$p_1 v_1 + p_2 v_2 = \hat{p}_1\hat{v}_1 + \hat{p}_2\hat{v}_2.$$

To this we add the identity $v_1 m_1 \hat{v}_1 + v_2 m_2 \hat{v}_2 = \hat{v}_1 m_1 v_1 + \hat{v}_2 m_2 v_2$. Writing it as

$$p_1\hat{v}_1 + p_2\hat{v}_2 = \hat{p}_1 v_1 + \hat{p}_2 v_2,$$

we obtain, after straightforward algebraic manipulation, the relation

$$(p_1 - \hat{p}_1)(v_1 + \hat{v}_1) + (p_2 - \hat{p}_2)(v_2 + \hat{v}_2) = 0.$$

Using the momentum equation in the form

$$p_1 + p_2 = \hat{p}_1 + \hat{p}_2 \qquad \text{or} \qquad p_2 - \hat{p}_2 = \hat{p}_1 - p_1,$$

we then are led to the relation

$$(p_1 - \hat{p}_1)(v_1 - v_2 + \hat{v}_1 - \hat{v}_2) = 0.$$

The first factor leads to the root $p_1 = \hat{p}_1$. From the second factor we may derive the root $p_1 = \check{p}_1$ after setting

$$v_1 = p_1/m_1, \qquad v_2 = (p_0 - p_1)/m_2.$$

In other words, $\check{p}_1$ can be obtained from the "velocity relation"

$$\check{v}_1 - \check{v}_2 = -(\hat{v}_1 - \hat{v}_2),$$

which is equivalent to the vanishing of the second factor. Once $\check{p}_1$ is found, $\check{p}_2 = p_0 - \check{p}_1$ can immediately be determined.

This velocity relation shows that, after colliding, the particles move away from each other, and that the solution $p_1 = \check{p}_1$, $p_2 = \check{p}_2$ therefore gives the outgoing momenta.

In our description of the process of impact we have made two assumptions which we now want to emphasize: The two particles hit each other and then bounce back from each other. Suppose the particle (2) is to the left of particle (1) before impact. Then particle (2) will catch up with particle (1) only if

$$\hat{v}_2 > \hat{v}_1,$$

and it will bounce back from it (and not go through it) if it remains on the left side of (1). This is possible only if the reverse inequality holds for the outgoing velocities:

$$\check{v}_2 < \check{v}_1.$$

Now, the first inequality together with the velocity relation $\check{v}_1 - \check{v}_2 = -(\hat{v}_1 - \hat{v}_2)$ implies the second one, and it is precisely this velocity relation that gave us our second solution, $p_1 = \check{p}_1,\ p_2 = \check{p}_2$.

Finally, we mention that it is not possible for the straight line to be tangent to the ellipse representing possible momenta after impact; for, if it were, the two roots would coalesce, that is

$$\check{p}_1 = \hat{p}_1, \qquad \check{p}_2 = \hat{p}_2,$$

and hence we would have

$$\check{v}_1 - \check{v}_2 = \hat{v}_1 - \hat{v}_2.$$

The velocity relation would then yield

$$\hat{v}_1 = \hat{v}_2,$$

and this would mean that the two particles have the same velocity initially. Thus they would never meet. But this contradicts the assumption that they hit each other at some time, hence the roots cannot coalesce.

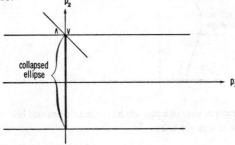

Figure 26. Graph in case one particle has mass zero and is at rest. Energy ellipse has collapsed into a segment

To illuminate the implications of what we have discussed we consider two special cases, namely the cases in which the mass of one of the particles is (a) very small and (b) very large when compared with that of the other one. For simplicity we describe these cases approximately by making the unrealistic assumption that one of the masses is (a) zero and (b) infinity, respectively.

(a) Accordingly we first assume that one of the masses is zero, $m_1 = 0$, say. Then the ellipse reduces to the segment

$$p_1 = 0, \qquad |p_2| \leq \sqrt{2m_2e_0}.$$

Suppose now particle (1) is to the right and at rest, i.e. $\hat{v}_1 = 0$, and particle (2) comes from the left with a velocity $\hat{v}_2 > 0$. Then clearly

$$p_0 = \hat{p}_2 = m_2\hat{v}_2 \quad \text{and} \quad 2e_0 = m_2\hat{v}_2^2.$$

After impact we have $\check{p}_1 = m_1\check{v}_1 = 0$ because $m_1 = 0$, whence

$$\check{p}_2 = p_0 - \check{p}_1 = p_0 = \hat{p}_2 \qquad \text{so that} \qquad \check{v}_2 = \hat{v}_2;$$

see Figure 26 and its caption. From the velocity relation one then derives

$$\check{v}_1 = 2\hat{v}_2.$$

Thus, the "heavy" particle (2) continues without being disturbed, while the "light" one has acquired double the velocity of the heavy one.

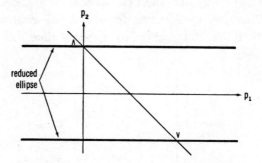

Figure 27. Graph in case one particle has infinite mass and is at rest. Energy ellipse reduces to a pair of straight lines

(b) Next assume $m_1 = \infty$; then the ellipse reduces to the pair of straight lines

$$p_2 = \pm\sqrt{2m_2e_0} ;$$

see Figure 27. This time we assume the heavy particle (1) to carry no momentum and thus to be at rest before impact, while the light particle (2) is to hit the heavy one from the left with a finite velocity $\hat{v}_2$. Since $\hat{p}_1 = 0$ we find $p_0 = \hat{p}_2 = m_2\hat{v}_2$ and $2e_0 = m_2\hat{v}_2^2$. Because $m_1 = \infty$ we have $\hat{v}_1 = \check{v}_1 = 0$. Therefore, the velocity relation yields

$\check{v}_2 = -\hat{v}_2$, so that the light particle reverses its velocity. Evidently the heavy one remains at rest. In that respect the situation is the same as in the first case. Note that the heavy particle, although it retains the mass $m_1 = \infty$ and the velocity $\check{v}_1 = 0$, acquires a finite momentum $\check{p}_1 = 2\hat{p}_2$ as can be seen from the momentum relation.

This peculiar result is of course due to the unrealistic assumption $m_1 = \infty$. If, instead, m_1 is taken very large but finite, the velocity $\check{v}_1$ will be approximately equal to $2m_2\hat{v}_2/m_1$ and thus will be very small, while the momentum $\check{p}_1 = 2m_2\hat{v}_2$ need not be small.

CHAPTER SIX

Inelastic Impact

In the process called "inelastic impact" two particles hit each other just as in elastic impact, but after impact they stick together and form a new particle which moves as a unit with a definite velocity $\bar{v}_0$. The mass m_0 of the new particle is the sum of the masses of the two original particles:

$$m_0 = m_1 + m_2.$$

It is a basic principle of mechanics that the law of conservation of momentum is valid also for inelastic impact (see Figure 28), that is

$$m_0\bar{v}_0 = m_1\hat{v}_1 +. m_2\hat{v}_2,$$

but the kinetic energy $\frac{1}{2}m \, | \, \bar{v}_0 \, |^2$ of the new particle is no longer equal to the sum of the kinetic energies $\frac{1}{2}m_1 \, | \, \hat{v}_1 \, |^2 + \frac{1}{2}m_2 \, | \, \hat{v}_2 \, |^2$ of the incoming ones.

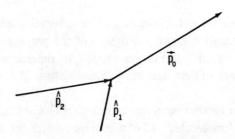

Figure 28. Momenta in inelastic impact

Note that now the velocity $\bar{v}_0$ of the new particle is completely determined by the velocities of the incoming ones; no additional information about the details of the impact process need be given. But what about the energies?

First of all we write down the formula

$$\tfrac{1}{2}m_1 \mid \hat{v}_1 \mid^2 + \tfrac{1}{2}m_2 \mid \hat{v}_2 \mid^2 \;=\; \tfrac{1}{2}m_0 \mid \bar{v}_0 \mid^2 + \tfrac{1}{2}(m_1 m_2/m_0) \mid \hat{v}_1 - \hat{v}_2 \mid^2,$$

which is easily verified by expressing $\bar{v}_0$ on the right hand side as $\bar{v}_0 = (m_1\hat{v}_1 + m_2\hat{v}_2)/m_0$ and then setting $m_0 = m_1 + m_2$. The two terms on the left are the kinetic energies of the incoming particles; the first term on the right is the kinetic energy of the compound particle. The last term will be called "excess energy". Note that this excess energy is never negative, and that it is zero only if $\hat{v}_1 = \hat{v}_2$, i.e. if the particles do not meet. In an actual impact, therefore, the excess energy is always positive. Consequently, the kinetic energy of the new particle is less than the sum of the kinetic energies of the incoming ones.

It will be observed that we did not say that the law of conservation of energy does not hold in inelastic impact. We left the possibility open that a form of this law different from the form it assumes for elastic impact might be valid. It is a remarkable fact of physics—if not the most remarkable one—that the notion of energy can be introduced in such a way that energy is conserved in all possible physical processes.

In contrast to what happens in elastic impact, the physical character of the particles changes in inelastic impact. This change is expressed by the presence of a new kind of energy after impact, referred to as "internal" energy. The law of conservation of energy is then maintained by setting this internal energy equal to the excess energy.

In order to describe the nature of this internal energy, one must give some information about the details of the process different from the kind of detailed information needed to determine the process of elastic impact. There are many possibilities. We select two as typical.

First we assume that each one of our particles actually consists of an assembly of "molecules", all of which move with the same velocity; moreover, we assume that during the process of impact the mechanism which has produced the uniformity of motion is disrupted and that the various molecules will eventually move in a random way, continually undergoing elastic impact with each other. The sum of all the kinetic energies of these motions, computed from velocities relative to the velocity $\bar{v}_0$, is the "internal" energy in this case. The requirement that this internal energy be identical with the excess energy $\frac{1}{2}(m_1 m_2/m_0) \mid \hat{v}_1 - \hat{v}_2 \mid^2$ is the form of the law of conservation of energy appropriate for the process described. The "random" motion involved is also referred to as "heat motion" and the corresponding internal energy as "heat energy". Of course, actual processes of inelastic impact in which heat is created are not as simple as the one described here.

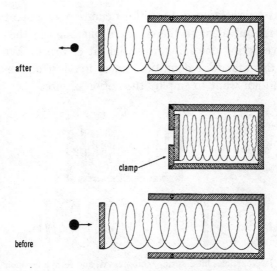

Figure 29. A particle compresses a spring and then is held in position

Before describing the other type of inelastic process we must mention the notion of "potential energy" in contrast to the notion of kinetic energy. For this purpose, we consider a spring neither stretched nor compressed, but held fixed at one end. We then assume that a particle moves towards the other end of the spring; see Figure 29. When the particle reaches the spring, it will start compressing it and, in doing so, will slow down and eventually lose all its velocity. At that moment we hold the spring in position (by a clamp, for example) and remove the particle. Now we place any other particle against the spring and release the spring (by removing the clamp). The new particle will gain speed while the spring expands, and when the spring reaches the unstretched position the particle will leave it and move away with constant velocity. The question is, what is this velocity? The answer is: the second particle moves with such a velocity that its kinetic energy is the same as that of the first particle before it hit the spring. If the mass of the second particle is less than that of the first particle its speed will be greater than that of the first one, and vice versa. Thus there is stored in the contracted spring the potentiality to yield the kinetic energy which was lost in the contraction process. In view of this fact one ascribes to the contracted spring a "potential energy" equal to the amount of kinetic energy it can yield.

It should be mentioned that the definition of potential energy given here is equivalent to the standard one, namely the work done by the force acting on the spring while it contracts. We have not employed this standard definition since, in view of our restricted aims, we did not want to employ the notion of force.

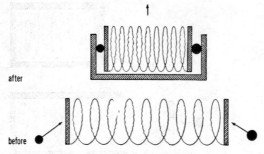

Figure 30. Two particles hit a spring, compress it, and form a compound particle

We now proceed to describe the second type of inelastic impact process we have selected for discussion. This process involves a primitive model. In it a spring is present *between the two particles* (it may be assumed attached to one of them), and it starts being compressed as soon as both particles have come in contact with it. While the spring is being compressed, the speed of each particle relative to the other, i.e. the relative speed $|\bar{v}_1 - \bar{v}_2|$, decreases. As soon as the relative speed has become zero, a lock snaps—we assume—so as to keep the compressed spring and the two particles in position. (See

Figure 30.) From then on the "compound particle" (i.e. the spring and the two particles) moves with the velocity $\bar{v}_0 = (m_1\bar{v}_1 + m_2\bar{v}_2)/m_0$. The mass of this new particle is the sum of the masses of the two constituents, provided that the mass of the spring is negligible. In this case the "internal energy", which equals the "excess energy" by virtue of the conservation law, is the potential energy stored in the compressed spring.

A compound particle of another type worth mentioning is a molecule that consists of two atoms held together by a chemical bond.

An *inelastic impact* can be run through in the *reverse* direction: a compound particle consisting of two constituents held together by a lock may break apart if the lock is opened. This will certainly happen if the two particles are separated by a compressed spring—or an equivalent mechanism—which is released at the time when the lock is opened. Such a process occurs in a gun and in a rocket. In order to shoot a light particle from a gun at a high speed one is forced to put up with the kick-back of the "heavy particle", the gun itself. In the case of a rocket, one wishes to impart speed to the heavy particle and is forced to abandon the light particle, the expensive fuel. For want of a more appropriate term we shall refer to an inelastic impact in reverse as an *explosion* process.

To be sure, such a process is again governed by the laws of conservation of momentum and energy. We denote the mass of the unexploded particle by m_0 and its velocity by $\bar{v}_0$. Then its momentum is

$$\bar{p}_0 = m_0\bar{v}_0,$$

but its energy is not simply the kinetic energy; instead, the energy

$$e_0 = \tfrac{1}{2}m_0\,|\,\bar{v}_0\,|^2 + g$$

of the unexploded particle consists of two parts: the kinetic energy $\tfrac{1}{2}m_0\,|\,\bar{v}_0\,|^2$ and the potential energy g stored in the mechanism. We assume that this stored energy is positive, $g > 0$. The two conservation laws then are

$$p_1 + p_2 = p_0, \qquad e_1 + e_2 = e_0.$$

In trying to determine the motion of the two particles ensuing after explosion we face the same difficulty we met in connection with the elastic impact: this motion is not completely determined by the laws of conservation; additional information is needed.

We make the same simplifying assumption that we made in discussing inelastic impact: we assume that both particles move in a straight line, the x-axis. The two conservation laws then assume the form

$$p_1 + p_2 = p_0, \qquad \frac{p_1^2}{m_1} + \frac{p_2^2}{m_2} = 2e_0,$$

or

$$m_1 v_1 + m_2 v_2 = m_0 v_0, \qquad m_1 v_1^2 + m_2 v_2^2 = 2e_0 = m_0 v_0^2 + 2g,$$

where p_1, p_2, p_0 are the x-components of the momenta of the three particles, m_1, m_2, $m_0 = m_1 + m_2$ their masses, and $v_1 = p_1/m_1$, $v_2 = p_2/m_2$, $v_0 = p_0/m_0$ their velocities.

The two solutions of this pair of equations are easily determined by virtue of the relation

$$m_1 v_1^2 + m_2 v_2^2 = \frac{(m_1 v_1 + m_2 v_2)^2}{m_0} + \frac{m_1 m_2 (v_1 - v_2)^2}{m_0}.$$

Identifying the stored potential energy g with the resulting excess energy

$$\frac{m_1 m_2 (v_1 - v_2)^2}{2m_0},$$

we find

$$v_1 - v_2 = \pm \sqrt{2g m_0 / m_1 m_2}.$$

Actually, we are looking for only one solution corresponding to a process in which the two particles move away from each other. Letting particle (1) be on the right of particle (2) we should have $v_2 < v_1$; therefore we must select the $+$ sign in the equation above:

$$v_1 - v_2 = \sqrt{2g m_0 / m_1 m_2}.$$

This equation together with

$$m_1 v_1 + m_2 v_2 = p_0 = m_0 v_0$$

allows us to find both v_1 and v_2. In particular, for $v_0 = 0$, we find

$$v_1 = \sqrt{2g m_2 / m_1 m_0}, \qquad v_2 = -\sqrt{2g m_1 / m_2 m_0}.$$

From the many questions that could be asked concerning such an explosion process, we select one. Suppose a rocket of a given mass m_1 carrying fuel and other substances of mass m_2 is to be fired from rest with a potential energy of a given amount g. One may then ask: for what value of the mass m_2 to be shot out (towards the left) is the velocity v_1 of the rocket as large as possible? To answer this question we investigate how the expression $v_1 = \sqrt{2gm_2/m_1m_0}$ varies when m_2 varies. If we let m_2 increase, the ratio

$$\frac{m_0}{m_2} = 1 + \frac{m_1}{m_2}$$

decreases, and hence v_1 increases; if we let m_2 tend to infinity, the ratio m_0/m_2 tends to 1 and hence v_1 approaches a finite limit,

$$v_1 \rightarrow \sqrt{2g/m_1} \quad \text{as} \quad m_2 \rightarrow \infty.$$

Therefore, v_1 can never exceed the value $\sqrt{2g/m_1}$, but can be brought arbitrarily close to this value by making m_2 sufficiently large. Thus we see that under the indicated circumstances it is advantageous to shoot out as much mass as possible. If the rocket is shot up vertically from the surface of the earth, one may, at the first instant, regard the fuel together with the whole earth as "the particle" shot out; infinity is evidently a good approximation for its mass.

The situation is, of course, different if the mass to be shot out must first be transported to the place where the explosion is to be set off. Moreover, in actual rocket design innumerable other factors must be taken into account to determine conditions of optimal performance. Still, a discussion of the type presented here is the first step in such an analysis.

A few remarks of a general character may be made here about our discussion of the problem of impact. Our presentation involved essentially only the mathematical features of the theory of impact and was not a realistic analysis of this process. Impact is a phenomenon of nature, and its theory is a branch of physics. This theory—just as any other theory of physics—was developed and verified in the course of time on the basis of long and varied experience. Without confronting such a theory with the evidence derived from actual experiences, it reduces to just a mathematical framework. In this sense we should say that we have presented here a mathematical framework for the theory of impact. While investigation of a mathematical framework as such is not physics, it may nevertheless be very helpful for physics, particularly if this framework involves concepts (such as the concept of vector in the case of impact) which make it clear cut and transparent.

It is a remarkable fact—far from obvious—that a very large part of physics permits itself to be fitted into a concise and elegant mathematical framework. In view of this fact, it is understandable that many physicists frequently let themselves be swayed into developing or accepting new formulations of physical theory by the sheer mathematical elegance of these formulations. It is again a very remarkable fact that ever so often they were right. To a large degree the theory of relativity is a case in point.

CHAPTER SEVEN

Space and Time Measurement in the Special Theory of Relativity

The special theory of relativity was originally concerned with the measurement of space and time by means of rigid rods and spring-driven clocks; and with the relationship of such measurements to observations made with the aid of electromagnetic waves. Later on, other propositions were developed in connection with this theory, such as modifications of the laws of conservation of momentum and energy in impact processes. It is the part of the theory concerned with impact that we intend to emphasize here; but we first must give a short account of the problem of space and time measurement.

Before one can measure distances in space one must adopt a "unit distance". To this end one should select two definite points and adopt the distance between them as the unit. Next one should employ a rigid rod whose length is just the unit distance, as verified by placing it between the two selected points. The distance between any two points may then be established by laying off the unit rod on the straight line connecting these points. To measure the time elapsed between two events one should proceed similarly with the aid of a spring that oscillates without external interference; such an oscillating spring will be called a "steady clock". In making these measurements it is taken for granted that it does not matter at what time the distance between the points is measured and at what place the time elapsed between the two events is observed.

To make clear what we mean by the "same place" at different times and the "same time" at different places, we assume that these places are marked on a definite body—such as the surface of the earth—called the "platform" in what follows. To verify that a time measurement is independent of the place on the platform where it is made, one should take a clock, synchronize it with a standard clock attached to a selected fixed point on the platform, carry it bodily to a different place on it, and compare it with a clock attached there. To verify that a distance measurement is independent of the time when it is made, one should repeat this measurement at a later time, using the same measuring rod.

It is one of the basic contentions of the "classical mechanics" of Galileo and Newton that the result of such measurements is independent of whether the platform is moving or at rest. It is here that the contentions of Einstein's mechanics differ from those of classical mechanics.

It is frequently stated that the differences between classical physics and the theory of relativity concern the nature of "space" and "time". We should like to emphasize that the questions we shall be discussing are not questions of philosophy but questions of physics. These questions concern specific physical structures—rigid bodies and oscillating elastic springs—and the behavior of these structures in space and time, but not the notions of space and time as such.

The difference in behavior of rigid rods and steady clocks as predicted by the two kinds of mechanics shows up, in particular, if one tries to find out—with the aid of rods and clocks—whether or not the reference platform is actually moving. Both classical and Einstein's kinematics* imply that it is not possible to decide this with the aid of rods and clocks alone. If other tools of time observation (such as light signals) were admitted, a decision could be reached provided rigid rods behaved as claimed in classical mechanics and light signals behaved as claimed in electromagnetic theory.

These theoretical claims were tested in 1887 by Michelson and Morley in a crucial experiment in which the surface of the earth served as a moving observation platform. Two rods of equal length were joined end to end at right angles; then light was sent out from the juncture and reflected from mirrors at the other ends of the rods.

*Kinematics here refers to that part of mechanics which is concerned with motion without regard to the mechanisms which produce or influence motion.

The difference of the times at which the signals arrived back at the starting point was measured, and it was found that there was no difference in re-arrival time. This contradicted the claims of classical physics since it was, of course, taken for granted that the earth actually moves.

Assuming that the speed of light c is independent of the motion of the sender—as claimed in electromagnetic theory—and that the lengths of the rods are not affected by their motion—as implied in classical mechanics—one readily finds the rearrival times $2ac/(c^2 - v^2)$ for a rod pointing in the direction of the motion of the earth, and $2a/\sqrt{c^2 - v^2}$ for a rod in a perpendicular direction. Here v is the speed of the earth at the point of observation and a the length of the rods. (See Figure 31.) Clearly these re-arrival times are different unless $v = 0$.

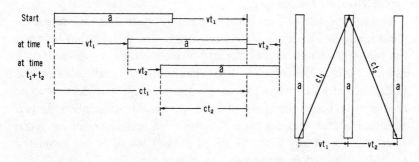

Bar in direction of earth's motion

$$ct_1 - vt_1 = a$$
$$ct_2 + vt_2 = a$$
$$t = t_1 + t_2 = 2ac/(c^2 - v^2)$$

Perpendicular bar

$$(ct_1)^2 - (vt_1)^2 = a^2$$
$$(ct_2)^2 - (vt_2)^2 = a^2$$
$$t = t_1 + t_2 = 2a/\sqrt{c^2 - v^2}$$

Figure 31. Computation of rearrival time in Michelson experiment

How to compute these re-arrival times is indicated in Figure 31. The time it takes the light signal to arrive at the mirror is denoted by t_1, while t_2 denotes the time it takes the signal to return to the starting point on the bar. The signal sent out in the direction of the motion of the earth has travelled the distance ct_1 during the time t_1; this length can also be described as the distance between the original initial point of the bar and its endpoint after the time t_1 and thus as $a + v_1 t_1$, since $v_1 t_1$ is the distance the endpoint has travelled during the time t_1. Therefore $ct_1 = a + v_1 t_1$. Similarly one derives the relation $ct_2 = a - v_2t_2$ for the return trip. For the signal sent out in a direction perpendicular to the motion of the earth the relations between ct_1 and v_1t_1 and between ct_2 and v_2t_2 are derived by means of the Pythagorean theorem.

The absence of a difference in the re-arrival times in Michelson's and Morley's experiment indicated that there was something wrong with classical mechanics provided, as was assumed with good reason, that the trouble did not stem from electromagnetic theory.

The celebrated Michelson–Morley discovery was the main starting point for the development of the "theory of relativity". The kinematics of this theory was developed so as to imply that it is impossible to decide whether or not a body moves at a given time, even if light signals or any other electromagnetic processes are used. A number of other rather spectacular claims were made by this theory and will be described in this chapter.

The significance of the claims concerning space and time measurement made in Einstein's kinematics lies in the relationship between the behavior of rods and clocks and of electromagnetic devices such as light signals. Nevertheless, the behavior of rods and clocks can be described without reference to electromagnetic processes—except that one quantity, the speed of light, enters this description. It is a remarkable fact that such an "intrinsic" description can be given simply and naturally with the aid of the notions of vector geometry suitably extended from "space" to "space-time".

Since every event in nature takes place somewhere in space and at a definite time, it is not at all unnatural to introduce an entity which is characterized by a point in space and by a definite time. Each such entity—customarily, though somewhat improperly, called "event"— may be characterized by four numbers: the three space coordinates x_1, x_2, x_3 of a point on a platform and the time t, i.e., the length of time passed since an arbitrarily chosen initial time. For this reason the manifold of these events is regarded as "four dimensional".

Instead of the time t we shall employ the related quantity $x_0 = ct$, where c is a constant, in fact the speed of light. The quantity x_0 is then the distance a particle would run through during the time t if it were moving with speed c.

We shall use the notation $\tilde{x}$ for an event and write

$$\tilde{x} = (x_0, x_1, x_2, x_3).$$

We use the same notation for the "event vector" which leads from the origin event

$$\tilde{O} = (0, 0, 0, 0)$$

to this point. The "coordinates" of the event are then the "components" of the event vector. We further introduce the "space component"

$$\bar{x} = (x_1, x_2, x_3)$$

of $\tilde{x}$ and also write

$$\tilde{x} = \{x_0, \bar{x}\}.$$

Finally, we introduce unit event vectors (omitting the tilde $\sim$)

$$u^{(0)} = (1, 0, 0, 0), \qquad u^{(1)} = (0, 1, 0, 0),$$

$$u^{(2)} = (0, 0, 1, 0), \qquad u^{(3)} = (0, 0, 0, 1),$$

so that we may write

$$\tilde{x} = x_0 u^{(0)} + x_1 u^{(1)} + x_2 u^{(2)} + x_3 u^{(3)}.$$

We note, in particular, that the event vector $x_0 u^{(0)} = ctu^{(0)}$ corresponds to the event taking place at the space origin $\bar{O} = (0, 0, 0)$ at the time t.

The assignment of the four coordinates to an event should be made with the aid of measuring operations of the kind described before. The measuring rod should be tested by comparing it with the three unit distances, the distances of the endpoints of the vectors $u^{(1)}$, $u^{(2)}$, $u^{(3)}$ from the origin $\bar{O}$. The clock should be tested by comparing it with a standard clock placed at the origin $\tilde{O}$.

In carrying out the measuring operations one should make sure that the measuring rod and the clock are at rest relative to the platform when they are compared with the standards and also when the final measurement is made. This remark is important since the deviations of Einstein's from classical kinematics arise just when a moving rod, or a moving clock, is compared with a resting rod or clock.

To describe the behavior of rods and clocks according to Einstein we suppose that a second platform is present which moves with a constant velocity relative to the first one. Again one can introduce a standard length, a standard clock, and coordinates in the manner described. These new coordinates will be denoted by $\bar{x}_0$, $\bar{x}_1$, $\bar{x}_2$, $\bar{x}_3$.

The units of length and time on the second platform should be the same as on the first one. To insure this one may assume that both platforms were at rest relative to each other at some earlier time so that the units could be identified. After having been accelerated temporarily, the second platform then should have acquired its final velocity. For simplicity, we assume that the origin event for the second platform is the same as that of the first one, $\tilde{O} = (0, 0, 0, 0)$.

Note that we must distinguish between the origin event $\tilde{O}$ and the origin space points $\bar{O}$ and $\bar{O}$ attached to the original and second platforms. Clearly, instead of saying that the two platforms have the same origin "event" $\tilde{O}$, we can also say that their origin "space points" $\bar{O}$ and $\bar{O}$ coincide at the time $t = 0$.

Now consider any event and ask for its coordinates (x_0, x_1, x_2, x_3) and $(\bar{x}_0, \bar{x}_1, \bar{x}_2, \bar{x}_3)$ with respect to the first and second platforms. According to the principles of classical kinematics the two time coordinates should be the same; they should be independent of whether the observation platform is at rest or in motion. The space vectors $\bar{x} = (x_1, x_2, x_3)$ and $\bar{x} = (\bar{x}_1, \bar{x}_2, \bar{x}_3)$ should differ only by the vector leading from the first origin $\bar{O}$ to the location of the second origin $\bar{O}$ at the time of observation.

The situation is quite different according to the principles of kinematics of Einstein. In particular, according to these new principles, the time coordinate $x_0 = ct$ is no longer independent of the motion of the observer; the time when an event takes place, as determined by a measurement, is claimed to depend on the velocity of the observer. Moreover, it is claimed that two different events, (1) and (2), which are found to occur at the same time $(t_1 = t_2)$ when observed from one platform, are found to occur at different times $(\bar{t}_1 \neq \bar{t}_2)$ when observed from the other platform.

These and other peculiarities claimed by Einstein in his kinematics can be deduced from a single law of his theory, namely the law that stipulates how the coordinates $\bar{x}_0, \bar{x}_1, \bar{x}_2, \bar{x}_3$ with reference to the second platform are connected with the coordinates x_0, x_1, x_2, x_3 with reference to the first one. This transformation law, in turn, can be deduced from another law involving a certain "inner product" of two event vectors. This new inner product is similar to that of four dimensional geometry, as described in Chapter 4, but differs in one important feature. All this we shall explain in the present chapter.

It would be possible to build up the formulation of Einstein's kinematics gradually, motivating its peculiar features more or less in the same way they have developed historically. One would start discussing basic experiments, such as that of Michelson, and deduce from this discussion how space and time measurements depend on the motion of an observation platform in strict analogy to the dependence of the measurement of electromagnetic quantities on the platform motion. Thus the interplay between mechanics and electromagnetics would be an essential ingredient of this approach.

Another approach to describing Einstein's kinematics would consist in first formulating it in a self-contained way and then drawing conclusions from it which could be confronted with the results of experiments and related to other fields of physics.

To gain a proper understanding one should go through both procedures. Most presentations of the theory of relativity emphasize the first approach. In the present exposition we follow the second one.

Suppose two event vectors $\bar{x}, \bar{y}$ are given as

$$\bar{x} = (x_0, x_1, x_2, x_3), \qquad \bar{y} = (y_0, y_1, y_2, y_3)$$

with respect to the original observation platform, while their components with respect to another platform are $\bar{x}_0, \bar{x}_1, \bar{x}_2, \bar{x}_3$ and $\bar{y}_0, \bar{y}_1, \bar{y}_2, \bar{y}_3$. Then the law referred to above as involving an "inner product" can simply be formulated through the relation

$$-y_0 x_0 + y_1 x_1 + y_2 x_2 + y_3 x_3 = -\bar{y}_0 \bar{x}_0 + \bar{y}_1 \bar{x}_1 + \bar{y}_2 \bar{x}_2 + \bar{y}_3 \bar{x}_3.$$

Inasmuch as the coordinates of the two events are determined by measurements made with the aid of rigid rods and steady clocks, it is clear that the law expressed by this relation refers to the behavior of such rods and clocks in motion. In fact, this law may be regarded as the *main contention of Einstein's kinematics*.

The above relation shows that the value of the expression on either side does not depend on the choice of the observation platform. It is also seen that this expression is formed from the components of the two event vectors $\bar{x}$ and $\bar{y}$ in almost the same way as the inner product of two three dimensional vectors is formed from their components with respect to a coordinate system; it differs only in the presence of the minus sign in front of the first term. It is therefore natural to call this expression an "inner product"

$$\bar{y} \cdot \bar{x} = -y_0 x_0 + y_1 x_1 + y_2 x_2 + y_3 x_3,$$

of the event vectors $\tilde{x}$ and $\tilde{y}$ and regard Einstein's kinematics as geometry in a four dimensional space governed by this peculiar inner product. This geometry is named after Minkowski, who developed it.

Before discussing the behavior of moving rods and clocks as implied by Einstein's law formulated above, we shall derive a few mathematical facts of Minkowski's geometry. In doing this we shall on occasion refer to moving or resting platforms where, mathematically speaking, we should refer to coordinate systems in four-dimensional space.

It may be expected that all of the statements of Euclidean vector geometry, such as those described in Chapter 4, have their counterparts in Minkowski's geometry. This is by and large the case; but there are a few vital differences. The origin of these differences is the minus sign in the expression for the Minkowskian inner product.

First of all, the basic properties of unit vectors can be expressed in terms of the inner products by the formulas

$$u^{(0)} \cdot u^{(0)} = -1,$$

(*) $$u^{(1)} \cdot u^{(1)} = u^{(2)} \cdot u^{(2)} = u^{(3)} \cdot u^{(3)} = 1,$$

$$u^{(0)} \cdot u^{(1)} = u^{(0)} \cdot u^{(2)} = u^{(0)} \cdot u^{(3)} = 0,$$

$$u^{(1)} \cdot u^{(2)} = u^{(1)} \cdot u^{(3)} = u^{(2)} \cdot u^{(3)} = 0,$$

which are immediately verified from the definition of these unit vectors in terms of their components; see page 67. Here only the first formula differs from the corresponding one in Euclidean geometry where the minus sign is absent.

Next we note that the components of the event vector

$$\tilde{x} = x_0 u^{(0)} + x_1 u^{(1)} + x_2 u^{(2)} + x_3 u^{(3)}$$

can be expressed as inner products of this vector and the unit vectors by the formulas

$$x_0 = -u^{(0)} \cdot \tilde{x}, \qquad x_1 = u^{(1)} \cdot \tilde{x},$$

$$x_2 = u^{(2)} \cdot \tilde{x}, \qquad x_3 = u^{(3)} \cdot \tilde{x},$$

which are readily verified. Also, the components of the barred unit vectors $\bar{u}$ with respect to the original platform are given as the inner products $-u^{(0)} \cdot \bar{u}, \quad u^{(1)} \cdot \bar{u}, \quad u^{(2)} \cdot \bar{u}, \quad u^{(3)} \cdot \bar{u};$ see Figure 32.

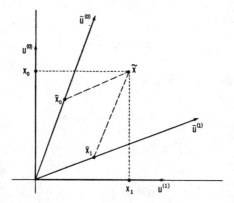

Figure 32. Unit vectors with respect to resting and moving platforms in Minkowski's geometry, and coordinates of an event with respect to both systems (here $\bar{u}^{(2)} = u^{(2)}$, $\bar{u}^{(3)} = u^{(3)}$ is assumed)

By way of an example, we consider the unit vectors associated with a platform moving with the velocity $\bar{v}$ in the x_1-direction relative to the original one and determine their components with respect to this "resting" one. The space origin $\bar{O}$ of the moving platform, which coincided at the time $t = 0$ with the origin $\bar{O}$ of the resting platform, will at the time t be at the point with the coordinates $x_1 = vt$, $x_2 = 0$, $x_3 = 0$ with respect to the resting platform. The corresponding event vector therefore is given by $\bar{x} = (ct, vt, 0, 0)$. As observed from the moving platform, its own origin $\bar{O}$ has the space coordinates $\bar{x}_1 = \bar{x}_2 = \bar{x}_3 = 0$. Hence the event vector associated with the point $\bar{O}$ is a multiple of the unit vector $\bar{u}^{(0)}$ of the moving platform whose barred components are $(1, 0, 0, 0)$. In other words, there is a number λ such that

$$(ct, vt, 0, 0) = \lambda \bar{u}^{(0)}.$$

(Unless otherwise stated, the components of a vector written in the form (, , ,) refer to the resting platform.) Taking the dot product of the vector $\lambda \bar{u}^{(0)}$ with itself, we obtain

$$(ct, vt, 0, 0) \cdot (ct, vt, 0, 0) = \lambda \bar{u}^{(0)} \cdot \lambda \bar{u}^{(0)} = \lambda^2 \bar{u}^{(0)} \cdot \bar{u}^{(0)}.$$

By using the definition of dot product in the left member of this equality and, in the right member, the relation $u^{(0)} \cdot u^{(0)} = -1$ (the first one in the array (*)), we find

$$v^2 t^2 - c^2 t^2 = -\lambda^2.$$

Hence, we obtain

$$\lambda = \sqrt{c_.^2 - v^2}\, t.$$

We denote the expression $1/\sqrt{1 - v^2/c^2}$ by β,

$$\beta = \frac{1}{\sqrt{1 - v^2/c^2}},$$

and write

$$\bar{u}^{(0)} = \left(\beta, \beta\,\frac{v}{c}, 0, 0\right).$$

The other conditions in the array (*), for the $\bar{u}$ in place of the u, are satisfied by

$$\bar{u}^{(1)} = \left(\beta\,\frac{v}{c}, \beta, 0, 0\right), \qquad \bar{u}^{(2)} = (0, 0, 1, 0), \qquad \bar{u}^{(3)} = (0, 0, 0, 1)$$

as can be readily verified. The components of the barred unit vectors given here, of course, refer to the resting platform.

We interrupt our mathematical analysis to discuss the significance of the last results for Einstein's kinematics. Suppose a platform moves as described, with velocity $\bar{v}$ in the x_1-direction when observed from a first platform. Consider the event given by the origin $\bar{O}$ of the moving platform at the time $\bar{t}$ as measured from the moving platform. Just as the event given by the original space origin $\bar{O}$ and the time t corresponds to the event vector $ctu^{(0)}$ (see p. 67), the event considered now corresponds to the event vector

$$c\bar{t}\bar{u}^{(0)}.$$

Since $\bar{u}^{(0)}$ is given by

$$\left(\beta, \beta\,\frac{v}{c}, 0, 0\right),$$

this vector is given by

$$(c\bar{t}\beta, \bar{t}\beta v, 0, 0)$$

in terms of components with respect to the original platform. Let t be the time of this event as measured from the original platform.

Then, as shown above, the corresponding event vector is given by

$$(ct, vt, 0, 0)$$

with respect to the original platform. Thus the components of the event vector considered are expressed in two different ways. Equating the corresponding expressions, we find

$$c\bar{t}\beta = ct \quad \text{and} \quad \bar{t}\beta v = vt.$$

That is to say, we obtain the relation

$$t = \beta\bar{t}.$$

This relation shows that *different times are ascribed to the same event when observed* with the aid of synchronized clocks *from differently moving platforms.*

Similarly, consider the event which corresponds to the point $\bar{x}_1 = \bar{a}$, $\bar{x}_2 = 0$, $\bar{x}_3 = 0$ and the time $\bar{t} = 0$ with reference to the moving platform. The corresponding event vector is evidently given by $\bar{a}\bar{u}^{(1)}$ and hence by

$$\left(\bar{a}\beta \frac{v}{c}, \bar{a}\beta, 0, 0\right)$$

in terms of the original coordinates. In other words, this event is ascribed to the time $t = \bar{a}\beta v/c$ and to the space point $x_1 = \bar{a}\beta$, $x_2 = x_3 = 0$ by an observer on the original platform who uses the same kind of rod and clock as the observer on the moving platform. Setting

$$a = \beta\bar{a}$$

we may say that the event considered, which has distance $|\bar{a}|$ from the origin when observed from the moving platform, has the distance $|a| = \beta |\bar{a}|$ when observed from the original one. Thus we see that *different distances are ascribed to a pair of points when observed from differently moving platforms.*

If one is willing to accept Einstein's kinematics—and experimental evidence is so strong that every physicist is willing to accept it—one is forced to accept the statements we have made as facts verifiable by observation.

Let us return to our mathematical analysis and state quite generally: once the motion of one (barred) platform with respect to another (unbarred) one is known, the unbarred components of the barred unit vectors and the inner products $u \cdot \bar{u}$ can be determined. Knowing these inner products one can express the unit vectors $\bar{u}$ in terms of the unit vectors u.

Suppose now an event vector $\bar{x}$ is given in terms of its barred components

$$\bar{x} = \bar{x}_0 \bar{u}^{(0)} + \bar{x}_1 \bar{u}^{(1)} + \bar{x}_2 \bar{u}^{(2)} + \bar{x}_3 \bar{u}^{(3)}.$$

Since the vectors $\bar{u}$ can be expressed in terms of the vectors u, the vector $\bar{x}$ can also be expressed in terms of the unit vectors u. Carrying this out one will find $\bar{x}$ as a linear combination of the vectors u. Identifying the coefficients in this combination with the unbarred components of $\bar{x}$, we find how the unbarred coordinates can be expressed in terms of the barred ones. Vice versa, one can find how the barred coordinates can be expressed in terms of the unbarred ones.

The resulting transformation formulas are named after H. A. Lorentz, who had recognized the importance of related formulas in electrodynamics before Einstein gave them his kinematic interpretation. We do not intend to describe these formulas in detail here. We mention, though, that they contain nothing that would allow one to distinguish a moving from a resting platform. This confirms the statement made earlier that, in Einstein's kinematics, such a distinction cannot be made with the aid of rods and clocks, just as it cannot be made in classical kinematics.

We now describe some of those features of Einstein's kinematics in which it differs from classical kinematics or, what is equivalent, in which Minkowski's geometry differs from Euclidean geometry.

Clearly, the most striking such feature is the minus sign in front of the term $y_0 x_0$ in the expression of the inner product. What is the significance of this sign? If the term $y_0 x_0$ were supplied with a plus sign, the resulting geometry would be nothing but the four dimensional analogue of Euclidean geometry. As was indicated at the end of Chapter 4, four dimensional Euclidean geometry does not differ in its basic features from three dimensional Euclidean geometry. On the other hand, as was said before, the Minkowski geometry— characterized by the minus sign in front of $y_0 x_0$—has features wherein it differs decisively from Euclidean geometry.

In particular, it is not possible to define the magnitude of every vector $\bar{x}$ as the square root of the inner product of this vector with itself, because for some vectors this inner product is negative. If $x \cdot x = \bar{x} \cdot \bar{x} - x_0^2$ is positive we may define the magnitude of the vector x by $|x| = \sqrt{x \cdot x}$; but if $x \cdot x < 0$ we define its magnitude by $|x| = \sqrt{-x \cdot x}$. The formulas

$$|x|^2 = \bar{x} \cdot \bar{x} - x_0^2 \quad \text{for} \quad x \cdot x > 0,$$

$$|x|^2 = x_0^2 - \bar{x} \cdot \bar{x} \quad \text{for} \quad x \cdot x < 0$$

may be regarded as analogues of the Pythagorean theorem.

Note that the space unit vectors $u^{(1)}$, $u^{(2)}$, $u^{(3)}$ belong to the first class since $u^{(1)} \cdot u^{(1)} = u^{(2)} \cdot u^{(2)} = u^{(3)} \cdot u^{(3)} = 1$. Thus the magnitude of each of these vectors is one. The time unit vector, on the other hand, belongs to the second class since $u^{(0)} \cdot u^{(0)} = -1$. (Its magnitude is also one.) It is because of these facts that vectors x with $x \cdot x > 0$ (Figure 33) are called "space-like", and vectors x with $x \cdot x < 0$ are called "time-like". Note that there are two kinds of time-like vectors, "forward" time-like ones with $x_0 > 0$ and "backward" time-like ones with $x_0 < 0$.

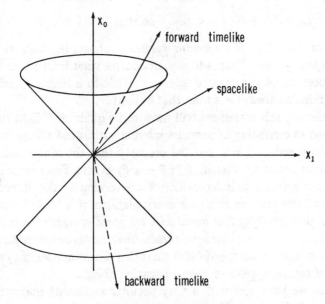

Figure 33. Spacelike and timelike vectors

These notions have been introduced because of their role in Einstein's kinematics. They are useful, in particular, in formulating the contention of this kinematics, supplementary to the "basic contention" formulated earlier, that the *time unit vector $u^{(0)}$ of any platform moving with constant velocity is always forward time-like.*

To illustrate the significance of this contention, let us consider the motion of a single particle with a constant velocity given by the space vector $\bar{v}$ relative to the original platform. For simplicity let us assume that the particle passes through the resting origin $\bar{O}$ at the time $t = 0$. At the time t this particle is then located at the point $\bar{x} = t\bar{v}$. The corresponding event is given by the vector $\{ct, t\bar{v}\}$ which we shall call "motion vector". Here we have used the abbreviation $\{x_0, \bar{x}\} = (x_0, x_1, x_2, x_3)$ introduced earlier. The contention that such a motion vector is always forward time-like for $t > 0$ is consistent with the statements made above about platforms since we may always consider a moving particle as the origin $\bar{O}$ of a moving platform. The event vector $\{ct, t\bar{v}\}$ of the moving particle at the time t can then be described as $t\bar{u}^{(0)}$, with $\bar{t} = t/\beta$ and $\bar{u}^{(0)}$ the time unit vector for the moving platform.

The statement that the vector $\{ct, t\bar{v}\}$ is time-like implies

$$-c^2 t^2 + t^2 \, |\, \bar{v}\, |^2 < 0, \qquad \text{so that} \qquad |\, \bar{v}\, |^2 < c^2.$$

Thus, *the speed $|\, \bar{v}\, |$ of the moving particle is always less than the speed c, the speed of light.* That is to say, no matter what mechanism might have been employed to accelerate a particle to a higher speed, this speed remains always less than that of light.

Evidently, this statement will have to be qualified if light itself is regarded as consisting of particles moving with speed c. Any particle with the speed $|\, \bar{v}\, | = c$ has the property that the inner product of its motion vector with itself, $t^2\, |\, \bar{v}\, |^2 - c^2 t^2$, is zero. Since the value of this inner product is independent of the motion of the observation platform, this relation holds for every platform if it holds for one. It follows that *anything that moves with the speed of light with respect to any platform moves with the same speed with respect to any other platform,* no matter with which speed ($<c$) this other platform itself may move. This, of course, applies to the propagation of light.

Again we have met with a very peculiar statement that must be accepted if Einstein's kinematics is accepted, and it is accepted by physicists since it agrees with the findings of Michelson and Morley.

The various facts discussed show clearly the major significance of the speed c for Einstein's kinematics. This speed c is supposed to be the speed of light. Now kinematics, as understood here, is concerned with the behavior of rigid rods and steady clocks; then what, one may wonder, has the speed of light to do with the behavior of rigid rods and steady clocks?

Before answering this question, a basic property of the propagation of light should be mentioned. This propagation is governed by the laws of electromagnetism formulated in the theory of Maxwell (about 1860). These laws have the remarkable implication that it is impossible to distinguish a moving from a resting platform by electromagnetic processes alone, provided space and time coordinates assigned to events on such platforms have just the properties postulated in Einstein's kinematics. If space-time measurement did not have these properties—for example, if it had the properties postulated in classical kinematics—it would be possible to distinguish between moving and resting platforms. How this could be done is explained at great length in detailed expositions of the theory of relativity and will not be explained here. As was said before, such considerations were the starting point for the theory of relativity, since experiments, such as that by Michelson and Morley, had strongly suggested that no observation allowed one to distinguish between a state of motion with constant velocity of a rigid body and the state of rest.

Still, one may wonder why the space-time behavior of rods and clocks should be in such complete harmony with the space-time properties of electromagnetic processes. To this one may say that, after all, a rigid rod is a body composed of particles held together by forces which are of electromagnetic origin, and the oscillations of a steady clock are regulated by forces of the same character. In view of this, one really should not be surprised that the space-time behavior of rods and clocks agrees with that of electromagnetic processes.

A few historical remarks may be added. Einstein formulated his theory in 1905 in a paper entitled *"Zur Elektrodynamik bewegter Körper"* ("On the Electrodynamics of Moving Bodies"). Before that time, many physicists and a few mathematicians were concerned with the relationship between kinematics and electrodynamics, and the fact that a number of them had made important contributions towards the theory of relativity should not be surprising. The physicist Lorentz had anticipated some of the basic facts and the mathematician Henri Poincaré, in his philosophical writings, discussed the dependence of measurements on the means of observation. In fact, at the same time as Einstein, in 1905, Poincaré published a paper on what he calls the "Postulate of Relativity"; in it he also formulates the "new mechanics" and analyzes the nature of the four dimensional coordinate transformation, which he calls "Lorentz transformation". Einstein in his work uses a somewhat less finished mathematical formalism than Poincaré, but he describes more effectively the significance of his "Principle of Relativity" for physical reality. He begins by pointing out with brilliant clarity that the notion of "same time" at "different places" is relative to the motion of an observer and describes in detail the role of measurements for the new kinematics. Thus, it was Einstein's work that initiated the development of the theory of relativity.

Among the numerous followers of Einstein we mention the mathematician Minkowski, who developed the geometrical interpretation of Einstein's kinematics, and the physicist Planck, who introduced those notions of force and momentum which are most properly adapted to Einstein's kinematics.

CHAPTER EIGHT

Momentum and Energy in the Special Theory of Relativity. Impact

Velocity is such an important notion of classical kinematics that one may naturally ask whether or not an analogous notion can be introduced in Einstein's kinematics as a (four dimensional) space-time vector. In a certain sense this is possible. While the ordinary velocity vector depends on the motion of the observation platform, the four dimensional analogue of this vector that we shall introduce is independent of this motion.

In the preceding chapter, in discussing the motion of a particle with the velocity $\bar{v}$ and passing through the origin O at the time $t = 0$, we introduced the "motion vector"

$$\{ct, t\bar{v}\}$$

which leads from the origin $\bar{O}$ to the event characterized by the time t and the place $t\bar{v}$ on the original platform where the particle is at the time t. To form an independent velocity vector we cannot just divide this motion vector by t, since the value of t depends on the motion of the platform. What is to be done? We simply divide the motion vector by that particular time t_E at which the event is observed from a platform that moves with the same velocity $\bar{v}$ as the particle, so that the particle is at rest when observed from it. Such a platform will be referred to as "special".

To compute the time t_E we note that, with reference to the special platform, the motion vector is given by

$$(ct_E, 0, 0, 0).$$

Thus, the event vector described by $\{ct, t\bar{v}\}$ with reference to the original platform has the components ct_E, 0, 0, 0 with reference to the special one. Equating the expressions for the inner product of this vector with itself, with reference to these two platforms, we find

$$c^2 t_E^2 = c^2 t^2 - t^2 \mid \bar{v} \mid^2,$$

whence

$$t = \beta t_E,$$

where

$$\beta = 1/\sqrt{1 - \mid \bar{v} \mid^2 / c^2}.$$

This expression for β is in agreement with the somewhat more special formula derived in the preceding section.

The value of t_E is called the "eigen-time" of the moving particle. Eigen-time, in partial translation of the German "Eigenzeit", means "time of its own". Evidently this value does not depend on what may be found when the particle is observed from another platform. To find an *analogue of the velocity vector* we divide the motion vector $\{ct, t\bar{v}\}$ by t_E; the vector thus obtained is described by

$$\bar{w} = \{\beta c, \beta \bar{v}\}$$

with reference to the original platform and by $(c, 0, 0, 0)$ with reference to the special one. We shall call this vector the "eigen-velocity vector". Note that each of its four components has the dimension of distance over time. Still, it is not strictly a velocity vector in the usual sense, since space and time are intermingled in it in a subtle way.

Our primary incentive in looking for an analogue of the ordinary velocity vector $\bar{v}$ was our wish to find an analogue of the ordinary momentum vector $\bar{p} = m\bar{v}$ which should enter the formulation of conservation laws governing impact processes. It is now easy to find an analogue that does not depend on the accidental choice of the observation platform. One simply employs the eigen-velocity vector $\bar{w}$ and considers the "four dimensional" momentum vector

$$\bar{p} = m\bar{w}$$

as the analogue of the "space" momentum vector; here m is the mass of the particle.

In terms of components with reference to the original platform, the vector $\bar{p} = \{p_0, \vec{p}\}$ is given by

$$\bar{p} = \{m\beta c, m\beta\vec{v}\}.$$

The space component of this vector is

$$\vec{p} = m\beta\vec{v}.$$

This relation looks like the classical relation between velocity and momentum except that the term $m\beta$ takes the place of the mass. Note that this "apparent mass" $m\beta$ depends on the velocity $\vec{v}$; it reduces to m if $\vec{v} = 0$, i.e. if the particle is at rest. Therefore, m is called the "rest mass". In the following we always mean the rest mass when we speak of the mass.

Let us consider the case that the speed $|\vec{v}|$ of the particle is small compared with c, the speed of light, so that β is approximately equal to 1. Then the space part of the new momentum vector $\vec{p}$ is approximately given as

$$\vec{p} \sim m\vec{v};$$

that is, it agrees approximately with the classical momentum.

The presence of a fourth component, the time-component

$$p_0 = m\beta c,$$

is a new feature of the new momentum vector. For small values of the speed $|\vec{v}|$ we have $\beta \sim 1$ so that p_0 is approximately equal to mc. The significance of the component p_0 will become clear only if one refines its approximate description.

We set

$$\beta = \frac{1}{\sqrt{1 - |\vec{v}|^2/c^2}} = \left(1 - \frac{|\vec{v}|^2}{c^2}\right)^{-1/2}$$

and use the binomial expansion

$$(1 + z)^\alpha = 1 + \alpha z + \cdots \quad \text{for} \quad \alpha = -\tfrac{1}{2} \quad \text{and} \quad z = -\frac{|\vec{v}|^2}{c^2}.$$

Retaining only the first two terms, we find the approximate expression

$$\beta \sim 1 + \frac{|\bar{v}|^2}{2c^2},$$

valid for small values of $|\bar{v}|/c$. Multiplying p_0 by c we obtain

$$cp_0 \sim mc^2 + \tfrac{1}{2}m\,|\bar{v}|^2.$$

Thus we see that the term cp_0 is approximately the classical kinetic energy $\tfrac{1}{2}m\,|\bar{v}|^2$ supplemented by an additional constant, mc^2. This is an important result.

Does this additional constant have any physical significance? At this moment we are not yet ready to ascribe such a significance to it. Nevertheless, it is convenient to give the term mc^2 a name at the present stage; we call it "rest energy". Note that the term mc^2 involves the rest mass and that the term cp_0 reduces to mc^2 for $|\bar{v}| = 0$.

The time-component p_0 of the vector $\bar{p} = \{p_0,\,\bar{p}\}$, multiplied by c, will be called its "energy component",

$$e_0 = cp_0,$$

or simply the "energy of the particle" and the vector $\bar{p}$ will therefore also be referred to as the "energy-momentum" vector. The term

$$cp_0 - mc^2 = m\left\{\frac{c^2}{\sqrt{1 - |\bar{v}|^2/c^2}} - c^2\right\}$$

will on occasion be referred to as the kinetic energy of the particle, since it agrees approximately with the classical kinetic energy if $|\bar{v}|/c$ is small. The three dimensional vector $\bar{p}$ will be called the "space momentum" of the particle. The vector $\bar{p}$ itself is more commonly referred to as the "energy-momentum vector".

The energy and the space momentum of a particle are related to each other by the formula

$$m^2c^2 = -\bar{p} \cdot \bar{p} = p_0^2 - |\bar{p}|^2,$$

which is easily verified. Clearly, this formula allows us to express the energy cp_0 in terms of the momentum $\bar{p}$:

$$p_0 = \sqrt{m^2c^2 + |\bar{p}|^2},$$

provided that $p_0 > 0$.

It is remarkable that the special theory of relativity enables one to ascribe an energy and a momentum to a particle even if its mass is zero. Of course, these quantities can then not be expressed in terms of mass and velocity; they must be defined in a different manner which we shall not describe here. The energy-momentum vector $\bar{p} = \{p_0, \vec{p}\}$ of such a massless particle should in any case satisfy the above relation which, for $m = 0$, becomes

$$p_0 = |\vec{p}|.$$

Particles without mass to which an energy-momentum vector with $m = 0$ can be ascribed do in fact occur in nature: "photons" or "light quanta" and "neutrinos" are such particles.

Certainly, it is tempting to employ the notions of energy and momentum in the manner here described. Of course, nothing will be achieved merely by introducing a tempting terminology. Actually, there is more to it. This terminology will suggest a new formulation of the laws of conservation of momentum and energy.

To describe such a new law we first consider the *elastic impact* of two particles, (1) and (2), which hit each other somewhere at some time. The laws governing the change of motion due to impact should of course be independent of the choice of the observation platform, and the space and time measures involved in it should be in accordance with the theory of relativity as described here. If the speeds of the particles are small compared with the speed of light, these laws should approximately be the ordinary laws of conservation of momentum and energy. It is now quite natural to assume that the *new impact law* just says that the *four dimensional momentum is conserved*.

To formulate this law we express the momenta $\bar{p}$ in terms of the eigen-velocity vectors $\bar{w} = c\bar{u}$ as $\bar{p} = m\bar{w}$, distinguish those of the first and second particles by superscripts (1) and (2) and mark the values of these quantities before and after impact with the symbols $\wedge$ and $\vee$. The masses of the two particles will be denoted by m_1 and m_2. Thus we write

$$\hat{p}^{(1)} = m_1\hat{w}^{(1)} \quad \text{and} \quad \hat{p}^{(2)} = m_2\hat{w}^{(2)} \quad \text{before impact;}$$

$$\check{p}^{(1)} = m_1\check{w}^{(1)} \quad \text{and} \quad \check{p}^{(2)} = m_2\check{w}^{(2)} \quad \text{after impact.}$$

The new impact law can then be written in the form

$$\hat{p}^{(1)} + \hat{p}^{(2)} = \check{p}^{(1)} + \check{p}^{(2)};$$

for, this relation evidently expresses the *conservation of* the four dimensional *momentum*. It is very natural to expect that this form of the impact law is the correct one. In fact, physicists accept this law in view of strong experimental evidence, especially in the field of elementary particles.

In terms of components with reference to any platform this law stipulates that the quantity

$$\frac{m_1 c^2}{\sqrt{1 - |\vec{v}^{(1)}|^2/c^2}} + \frac{m_2 c^2}{\sqrt{1 - |\vec{v}^{(2)}|^2/c^2}}$$

and the space vectors

$$\frac{m_1 \vec{v}^{(1)}}{\sqrt{1 - |\vec{v}^{(1)}|^2/c^2}} + \frac{m_2 \vec{v}^{(2)}}{\sqrt{1 - |\vec{v}^{(2)}|^2/c^2}}$$

are the same before and after impact. (Here we assume that the masses m_1, m_2 are not zero.) The last term represents the new "total space momentum", the first one is the new "total energy". It is an important fact that in the new formulation the laws of conservation of energy and momentum, when expressed without reference to a particular coordinate system, form an inseparable unit. We recall that the two laws of conservation for classical impact also form a unit inasmuch as, taken together, these two laws are independent of the motion of the observer, while each separately does depend on his motion. In the present theory this unity is achieved in a conceptually more direct way than in classical mechanics, simply by regarding energy (divided by c) and momentum as components of a single four dimensional vector. Before formulating the energy law for classical elastic impact we felt impelled to refer to the forces acting between the colliding particles. In the present theory the energy law is forced upon us by the momentum law through the requirement that these laws combine into one conservation law for space-time vectors involving only the masses and the motions of the two particles. The impact can then be characterized as elastic by the sole requirement that the masses of the particles remain unchanged, a requirement which would not be sufficient for this purpose in classical mechanics. That the requirement of unchanged masses is more restrictive in relativistic than in classical mechanics will become even more apparent in the study of inelastic impact.

The outcome of an impact process as predicted by the new conservation law evidently differs from that predicted by the old law; still, the law reduces approximately to the old laws of conservation of energy and momentum when the speeds $|\bar{v}^{(1)}|$ and $|\bar{v}^{(2)}|$ are small compared with the speed of light c.

Being derived from a few clear cut requirements, the new conservation law is quite compelling. Once formulated, it was naturally accepted as a basic law of nature. As said before, this conviction has been justified by a wealth of experimental evidence, in particular, by evidence concerning the interaction of elementary particles.

The discussion of the relationship between the motions before and after impact can be carried out just as in the classical case. Assuming the motions to proceed only in one direction, one may introduce the corresponding components of momentum $\bar{p}^{(1)} = (p_0^{(1)}, p_1^{(1)}, 0, 0)$ and $\bar{p}^{(2)} = (p_0^{(2)}, p_1^{(2)}, 0, 0)$ of the particles (1) and (2). The *law of conservation of momentum* then reads

$$p_1^{(1)} + p_1^{(2)} = p_1^{(0)},$$

as before, where $p_1^{(0)}$ is the total momentum of the incoming particles. The *law of conservation of energy*, however, becomes

$$\sqrt{m_1^2 c^2 + (p_1^{(1)})^2} + \sqrt{m_2^2 c^2 + (p_1^{(2)})^2} = e^{(0)}/c,$$

where $e^{(0)}$ is the incoming total energy. Thus, in terms of momenta, the new and old laws differ only in the relation expressing the "energy balance". The graph of the last equation in the $(p_1^{(1)}, p_1^{(2)})$-plane is an oval whose properties are easily discussed. The analysis of the momenta before and after impact can thus be carried out about as easily as for ordinary elastic impact. We shall not do this here.

We proceed to discuss the last item of this exposition, the *inelastic impact* in the theory of relativity. *It is in connection with the inelastic impact that the term mc² can be given a significant physical interpretation.* In fact, Einstein, in an appendix to his work of 1905, did consider a certain inelastic impact for this purpose.

Suppose then our two particles, which have the momenta $\bar{p}^{(1)}$ and $\bar{p}^{(2)}$ before they hit each other, stay together after impact and form a new particle with the momentum $\bar{p}^{(0)}$. Of course, we assume that the conservation law

$$\bar{p}^{(1)} + \bar{p}^{(2)} = \bar{p}^{(0)}$$

governs such a process. There is, however, a difficulty.

We recall that, in the case of classical inelastic impact, the law of conservation of energy held only after it had been modified to take into account the formation of an internal energy of one form or another, in addition to the kinetic energy of the compound particle. As we shall show presently, a similar difficulty arises here; however, the same device for overcoming it cannot be used in the framework of the theory of relativity because the energy component cp_0 depends on the choice of the observation platform, while the addition of internal to kinetic energy should be independent of such a choice.

Still there is a way of satisfying all conservation laws; this is achieved not by adding an internal energy to the energy component cp_0, but by postulating that the *rest mass of the compound particle changes appropriately*. The rest mass of the compound particle is then not just the sum of the rest masses of the two colliding particles.

To see how this postulate operates let us consider an observation platform with respect to which the compound particle is at rest, so that

$$\vec{p}^{(0)} = 0, \qquad p\,_0^{(0)} = m_0 c.$$

The last equation here is equivalent to the relation

$$\frac{m_1}{\sqrt{1 - |\vec{v}^{(1)}|^2/c^2}} + \frac{m_2}{\sqrt{1 - |\vec{v}^{(2)}|^2/c^2}} = m_0,$$

by means of which the rest mass m_0 of the compound particle is determined. If the process is observed from a different platform, the mass m_0 will enter the expressions for the components of the momentum of the compound particle.

The new rest mass m_0 will, in particular, affect the new rest energy $m_0 c^2$. From the last formula it is obvious that the new rest mass is greater than the sum of the rest masses of the constituents:

$$m_0 > m_1 + m_2.$$

The same is therefore true of the "rest energies":

$$m_0 c^2 > m_1 c^2 + m_2 c^2.$$

The excess of rest energy, $m_0 c^2 - m_1 c^2 - m_2 c^2$, takes the place of the excess energy in classical inelastic impact.

The description of the excess rest energy in terms of a change in rest mass does not preclude its interpretation as heat energy or potential energy of some sort or other. Under appropriate circumstances, such interpretations have been given explicitly. Of course, any change of internal energy is coupled with a change of rest mass. This being so, it is very tempting to assume—as Einstein did in 1905 —that the *rest energy consists completely of internal energy* of some sort or other which may be released as kinetic energy in an appropriate process.

To discuss this proposition we consider the inelastic impact in reverse, that is, the *explosion process*. If a compound particle splits into two, the total rest mass of the fragments will be less than the original rest mass, so that the decrease in rest energy supplies the needed kinetic energy of the particles flying apart:

$$m_0 c^2 - m_1 c^2 - m_2 c^2 = m_1 \left\{ \frac{c^2}{\sqrt{1 - |\vec{v}^{(1)}|^2/c^2}} - c^2 \right\}$$

$$+ m_2 \left\{ \frac{c^2}{\sqrt{1 - |\vec{v}^{(2)}|^2/c^2}} - c^2 \right\}.$$

If the explosion process consists in breaking the chemical bond that binds the atoms in a molecule, the decrease in rest energy will induce only a relatively small decrease in rest mass. Nuclear bonds, on the other hand, are, generally speaking, much stronger than chemical bonds. If such a bond breaks, the change in rest mass will be noticeable, and a sizeable fraction of the rest energy will be transformed into kinetic energy. This is what happens in a nuclear explosion.

Suppose one does not know the internal mechanism that holds the compound particle together. Then one can still say that the amount of internal energy which could be transformed into kinetic energy is at most equal to mc^2. Einstein's proposition implies the possibility that every particle is some sort of compound particle whose internal energy might be released.

Naturally, one wonders what would happen if all of the rest energy of the particle were transformed into kinetic energy through an explosion. In such a case both terms on the right-hand side in the formula above would be zero. Since the quantity in each bracket is positive, both m_1 and m_2 would be zero; that is to say, the fragments would be particles with mass zero.

We mentioned before that one may ascribe energy cp_0 and momentum $\vec{p}$ to such massless particles, but not velocity. The conservation relations for an explosion process in which two such particles are created,

$$\vec{p}^{(1)} + \vec{p}^{(2)} = 0, \qquad p_0^{(1)} + p_0^{(2)} = m_0c,$$

can well be satisfied; in view of $p_0 = |\vec{p}|$, we clearly have

$$p_0^{(1)} = p_0^{(2)} = \tfrac{1}{2}m_0c.$$

As was mentioned above, particles to which energy and momentum can be ascribed in this manner occur in nature: "photons" (i.e. particles of light) and "neutrinos" are such particles. Processes in which the rest energy is completely transformed into kinetic energy of massless particles (in a somewhat more complicated manner than in the process described above) have also been observed in nature, in particular in the transformation of matter into light.

This and related observations in nuclear physics justify Einstein's early contention that the rest energy mc^2 is an internal energy that may be released—partly or completely—in an explosion process. It was our ultimate aim in tracing the Pythagorean theorem through its various metamorphoses to lead up to this astounding fact, expressed in the famous formula

$$e = mc^2.$$